CW01202649

EXS 48:
Experientia Supplementum,
Vol. 48

Birkhäuser Verlag
Basel · Boston · Stuttgart

Azospirillum II
Genetics, Physiology, Ecology

Second Workshop held at the University
of Bayreuth, Germany
September 6–7, 1983

Edited by
W. Klingmüller

1983

Birkhäuser Verlag
Basel · Boston · Stuttgart

The workshop was organized by the University of Bayreuth and sponsored by
BASF Ludwigshafen, Ciba-Geigy Basel, Universitätsverein Bayreuth,
Bayrisches Staatsministerium für Landwirtschaft und Forsten.

Volume Editor
Prof. Dr. W. Klingmüller
Lehrstuhl für Genetik
Universität Bayreuth
Universitätsstrasse 30
Postfach 3008
D-8580 Bayreuth (FRG)

CIP-Kurztitelaufnahme der Deutschen Bibliothek

Azospirillum : genetics, physiology, ecology ;
workshop / organized by the Univ. of Bayreuth
and sponsored by BASF Ludwigshafen ... - Basel ;
Boston ; Stuttgart : Birkhäuser

2. Held at the University of Bayreuth, Germany,
September 6-7, 1983. - 1983.
   (Experientia : Suppl. ; Vol.48)
   ISBN 3-7643-1576-8
   [Erscheint: November 1983].

NE: Universität ‹Bayreuth›; Experientia /
Supplementum

All rights reserved.
No part of this publication may be reproduced, stored in a retrieval system, or transmitted
in any form or by any means, electronic, mechanical, photocopying, recording or
otherwise, without the prior permission of the copyright owner.

© 1983 Birkhäuser Verlag Basel
Printed in Switzerland by Birkhäuser AG, Graphisches Unternehmen, Basel
ISBN 3-7643-1576-8

PREFACE

W. Klingmüller

Lehrstuhl für Genetik, Universität Bayreuth,
Universitätsstraße 30, 8580 Bayreuth, FRG

On September 6th and 7th 1983 the second workshop on "Azospirillum: Genetics, Physiology, Ecology" took place at the University of Bayreuth, West Germany, organized by the genetics department. There were about 50 participants, who came from German research institutions, from other European countries, from Israel, Egypt and North and South America. The first such workshop had taken place two years ago in Bayreuth too, hence the organizers could draw on the experiences then obtained.

Azospirilla have, during the past ten years, found an ever increasing scientific interest, because first, these soil bacteria carry the genetic information for binding molecular nitrogen from the air, and second, they live in close vicinity to the roots of several grain crops and forage grasses. By exploitation of these two properties, it is hoped to develop inoculation procedures that result in yield increases in agriculture, in particular in soils poor in nitrogen.

The reports on the first afternoon focussed, as a result of the Bayreuth research interest, on the genetic basis and the regulation of nitrogen fixation in Azospirillum. Here, mainly by application of most modern gene technological approaches, considerable progress in the understanding of the situation has been made, and was documented in the corresponding reports.

Broadly discussed - on both days - were also metabolic properties, and those features of Azospirilla that seem important for their colonizing the roots of the host plants, e.g. pH dependence of growth, preferred utilization of carbon sources, hormonal effects, chemo- and aerotaxis.

On the following morning, a number of extensive reports brought data, supporting the contention that inoculation of plants, not only in the greenhouse, but also in field trials, can indeed result in growth respones, e.g. increased root development, shoot height, and yield in general. This was shown for wheat, sorghum and maize. These reports came from the Israel and Egypt groups, and up to now have bearing on limited regional soil and climatic conditions. However, they should enhance efforts of other groups, to develop inoculation methods for saving on mineral nitrogen fertilizers in the cultivation of grain crops. No wonder that among the participants, there were already members of two different industrial corporations from the fertilizer and the seed sectors.

The opportunity for an intense exchange of experiences and ideas was exploited not only during the scientific meetings, but also during social and informal events. Many scientific and personal contacts were established, and it is hoped that these will pay off in the future. The foreign participants were impressed by Bayreuth University, its research institutions, and in particular the genetics department. The pleasant city, and a splendid weather, added positively to the overall impression.

Thanks to the speakers, who had been asked to bring the manuscripts with them ready for print, and indeed did so, and by the good will and effectiveness of the publisher, it is possible, to bring these proceedings on the market within three months after the workshop. This is exceptional. It should help the participants to recapitulate and better understand the contributions. It should moreover help all other interested people, to judge the present state of Azospirillum work, and to evaluate the prospects of working with this organism in the future.

CONTENTS

Preface 8
    W. KLINGMUELLER

Ten Years Azospirillum 10
    J. DOEBEREINER

Genetic Analysis in Azospirillum 24
    M. BAZZICALUPO AND E. GALLORI

Recent Developments in the Genetics of Nitrogen Fixation in Azospirillum   29
    S.K. NAIR, P. JARA, B. QUIVIGER AND C. ELMERICH

Molecular Cloning of Nitrogen Fixation Genes from Azospirillum 39
    W. WENZEL, M. SINGH AND W. KLINGMUELLER

Site-Directed Transposon Mutagenesis of Cloned nif-Genes of Azospirillum
brasilense 47
    M. SINGH AND W. KLINGMUELLER

Uptake and Fate of Plasmid-DNA in Azospirillum lipoferum A23 56
    G. SCHWABE AND U. WEBER

Nif Mutants of Azospirillum brasilense: Evidence for a Nif a Type
Regulation 66
    F.O. PEDROSA AND M.G. YATES

Mutants of Azospirillum Affected in Nitrogen Fixation and Auxin Production   78
    A. HARTMANN, A. FUSSEDER AND W. KLINGMUELLER

Motility Changes in Azospirillum lipoferum 89
    T. HEULIN, P. WEINHARD AND J. BALANDREAU

Attraction of Azospirillum lipoferum by Media from Wheat-Azospirillum
Association 95
    D. HEINRICH AND D. HESS

Nitrogen Fixation and Denitrification by a Wheat-Azospirillum Association   100
    H. BOHTE, A. KRONENBERG, M.P. STEPHAN, W. ZIMMER AND G. NEUER

Effect of Oxygen Concentration on Electron Transport Components and
Microaerobic Properties of Azospirillum brasilense 115
    Y. OKON, I. NUR AND Y. HENIS

Nitrogen Fixation (C_2H_2-Reduction) and Growth of Pure and Mixed Cultures
of Azospirillum lipoferum, Klebsiella and Enterobacter sp. From Cereal
Roots in Liquid and Semisolid Media at Different Temperatures and Oxygen
Concentrations 127
    G. JAGNOW

Ecological Factors Affecting Survival and Activity of Azospirillum in
the Rhizosphere 138
    S.L. ALBRECHT, M.H. GASKINS, J.R. MILAM, S.C. SCHANK AND R.L. SMITH

Forage Grasses Inoculation with Gentamicine and Sulfaguanidine
Resistant Mutants of Azospirillum brasilense 149
    A. MAROCCO, M. BAZZICALUPO AND M. PERENZIN

Association between Wheat and Azospirillum lipoferum under Greenhouse
Conditions: Increase of Yield and Thousand Corn Weight 159
    TH. MERTENS AND D. HESS

Benefits of Azospirillum Inoculation on Wheat: Effects on Root Development,
Mineral Uptake, Nitrogen Fixation and Crop Yield 163
    Y. KAPULNIK AND Y. OKON

Contribution of Azospirillum SPP. to Asymbiotic N_2-Fixation in Soils and
on Roots of Plants Grown in Egypt 171
    N.A. HEGAZI

Resumé 190
List of Participants 193

TEN YEARS AZOSPIRILLUM

J. DÖBEREINER

Programa Nacional de Pesquisa em Biologia do Solo - EMBRAPA
Seropédica 23460, Rio de Janeiro, Brazil

Introduction

The first observations about the frequent occurrence of a vibrio-like organism with very characteristic movements which seemed to be associated with N_2 fixing root pieces were made only 10 years ago (J.J. Peña & J. Döbereiner 1973, unpublished). For more than a year this organism remained unrecognized because colonies reduced C_2H_2 only when mixed with contaminants, just like in the early experiments of Beijerinck (5) where only enrichment cultures fixed N_2 in liquid medium. Also Becking (4) observed $^{15}N_2$ incorporation only when yeast extract was added to the medium. The introduction of semisolid media was then the key to the identification of *Spirillum lipoferum* as true N_2 fixing microaerobe (16). In that paper, presented 1974 in Pullman several important observations were communicated which at the time seemed difficult to believe: (a) *Spirillum lipoferum* was identified in soils under grasses and several forms were distinguished; (b) These organisms are wide spread in soils, rhizosphere and roots; (c) Sensitivity to O_2 of N_2-fixation is the reason for the success of semi-solid media for isolation of these organisms; (d) Root tissues seem to be infected by these organisms; (e) Malate, the predominant organic acid in roots is the favored C source; (f) Plant-bacteria interactions exist; (g) N_2 fixation associated with grasses can reach economically important amounts.

Ten years later most of these observations have been confirmed and further expanded as very well exemplified in the present Workshop and also in several recent reviews (e.g. 38; 33; 30; 8; 31).

Taxonomy of Azospirillum spp.

There are now three species of Azospirillum the main characteristics being summarized in Table 1. In addition to the description by Tarrand et al. (36) a new species, A. amazonense, was proposed recently (26) which is recognized upon isolation by a deep diffuse pellicle in N-free semi-solid malate medium (NFb) which does not mouve to the surface and promotes little nitrogenase activity. This was shown to be due to the sensitivity to alcaline reaction of the new organism. In sucrose medium where there is little pH change, a thick surface pellicle is formed and nitrogenase activity becomes as high as in the other species. Other differencial characteristics are shown in Table 1. Electron micrographs are shown in Fig. 1.

Ecology and pH effects

The occurrence of all three species can be compared in Table 2. Although less abundant the new species, as the others, is strongly stimulated in the grass rhizosphere and also occurres in surface sterilized roots. Its role in associative N_2 fixation has not yet been accessed. It has also been isolated from roots of certain fast growing palm trees (Bactris grassipus) in the Amazon region and from cassava (one sample).

Earlier observations on pH requirements of A. brasilense strain Sp 7 (11; 29) and of the occurrence of Azospirillum spp in soils (18) have led to the general believe that these organisms like Azotobacter are very pH demanding. It was also observed that the incidence of Azospirillum in grass roots was much less dependent on the soil pH than that in soil (13). Recent observations indicate that not only A amazonense but also most of the A. lipoferum and some A. brasilense strains grow as well or better at pH 6.3 or even 5.7 as compared to pH 7.0 (Table 3). Especially strains isolated from roots seem to prefere acid media. This is interesting because the colonization of the organisms within xylem vessels as observed by (31) and (17) where the pH occilates between 5 and 6 (Fig. 2) has been

Table 1. Differences between *Azospirillum* spp (from 28)

	A. amazonense	A. lipoferum	A. brasilense
Growth on medium with pH above 6.8	very poor	good	good
Colony type on potato agar	white flat raised margin	pink raised	ping raised
Tolerance to O_2 for nitrogenase activity	very low	low	low
Dissimilation of			
$NO_3 \rightarrow NO_2$	$\underline{+}$[a]	+	+
$NO_2 \rightarrow N_2O$	−	$\underline{+}$	$\underline{+}$
Cell width (nm, N_2 grown)	0.68 ± 0.08	1.0 to 1.5[b]	0.9 ± 0.03
Polar flagellum	+	+	+
Lateral flagella on nutrient agar	−	+	+
Polymorph cells in alcaline media	−	+	−
Biotin requirement	−	+	−
Use of sucrose	+	−	−
Generation time for N_2 dependent growth at optimal pO_2	10h	5–6h	6h
DNA base comp. (mol % G + C)	67–68	69–70	69–70

[a] Explanation of signs: + positive in more than 90% of the strains
$\underline{+}$ positive in less than 50% of the strains
− negative

[b] Cells of *A. lipoferum* may become even wider and much longer in older alcaline cultures

Table 2. Occurrence of *Azospirillum* spp. associated with tropical forage grasses[a] (from 28)

	Soil	Washed	Surface steril. roots[b]
Total nº of samples	115	345	345
% samples containing A. *lipoferum* and/or A. *brasilense*	94	85	45
% samples containing A. *amazonense*[c]	42	53	32
% samples containing more than 10^5 A. *brasilense* and/or A. *lipoferum*	8	34	10
% samples containing more than 10^5 A. *amazonense*	2	16	1
% samples containing more A. *amazonense* than the two known species	37	27	20

[a] *Hyparrhenia rufa*, *Digitaria decumbens* cv. transvala, *Brachiaria decumbens* cv. pangola, *Pennisetum purpureum* cv. elefante, *P. purpureum* cv. cameron grown in a field experiment with 3 N levels and with 5 replicates, sampled monthly during one year. There were no apparent differences between N levels or plant genotypes.

[b] 15 min. in chloramine-t

[c] Estimated by the occurrence of low diffuse pellicles with marginal nitrogenase activity in vials for MPN counts in NFb medium originally thought to count only the known *Azospirillum* spp. In vials which contained the latter, A. *amazonense* would not have been recognised.

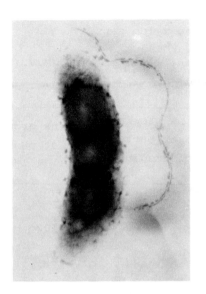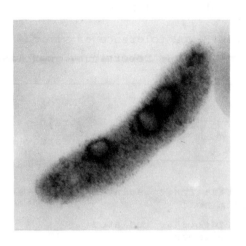

Fig. 1 - *A. amazonense* (A) in liquid NFb medium at pO_2 0.005 (B) in nutrient broth (35.560 x)

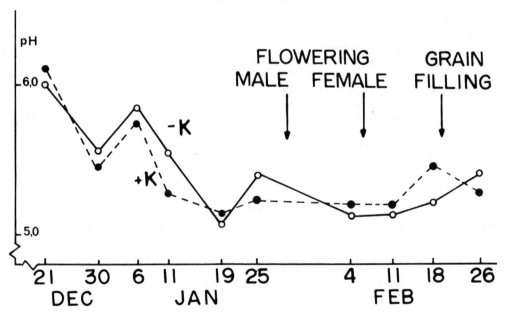

Fig. 2 - Changes of pH in xylem sap of field grown maize with and without K fertilizer

thought to be limited by the pH of the xylem sap.

Table 3. pH tolerance of various *Azospirillum* strains (M.V. Reis & J. Döbereiner, unpublished)

	Nº of Strains tested	Strains growing equal or better on acid medium[a] %		Strains growing best at pH 7,0 %
		pH 5,7	pH 6,3	
A. *brasilense* isolated from soil	9	0	0	100
A. *brasilense* isolated from roots	17	12	53	47
A. *lipoferum* isolated from soil	2	0	50	50
A. *lipoferum* isolated from roots	8	38	88	12
Atyp. A. *lipoferum* like Sp 242[b]	8	0	100	0

[a] Evaluated by N_2-ase activity in N free semisolid glycerol medium with pH 5,7, 6,3 or 7,0.

[b] Several strains when isolated from sterilized maize or sorghum roots grew well on glucose and mannitol but later upon storage lost their ability to grow on these two carbon substrates. All these strains are related by immunodiffusion tests and unrelated to other A. *brasilense* or A. *lipoferum* isolates from other sources.

It is well known that A. *lipoferum* strains turn into immobile aberrantly large involution forms in alcaline but not in neutral or acid media (36). Also A. *lipoferum* stock cultures are lost with much higher frequency than A. *brasilense* strains, both being stored on malate medium which turns alcaline. This indicates a certain sensitivity to alcaline reaction which is also the reason for the typical rugh *Azospirillum* type colonies on potatoe medium which contains malate. If malate is replaced by glucose large smooth white colonies are formed because the bacteria continue growing when the medium does not turn alcaline.

The understanding of these implications of the pH of the medium led to the isolation of A. amazonense which is still more sensitive to alcaline reaction. The typical white soucer colonies of this organism are due to its ability to use both malate and sucrose both present in the potatoe medium. As with A. lipoferum abundant slimy growth is observed when the malate is omitted from the medium.

Interference of pO_2

It was soon recognized that the key role of semi-solid media for Azospirillum isolation is not as originally thought a better imitation of root-soil interfaces, but rather is due to its physical conditions which permit migration of the organisms to the depthh where O_2 diffusion rates are in equilibrium with bacterial respiration rates. This has now been clearly established (40; 26) but is still not universally recognized. In the literature various oxygen concentrations either in the gas phase above or within the culture are claimed to the optimal for Azospirillum N_2 fixation (29; 39). There is however no such thing as an optimal pO_2 because it is dependent on the density of the culture, on the respiration rate of the organisms and on the shaking rate and liquid - gas interfaces. We therefore prefere to define the optimal pO_2 for N_2 dependent growth of a microaerobe as the maximal O_2 input rate which can be used up by respiration resulting in a pO_2 near zero at the cell surface.

Root Infection

Until 1970 "free living" N_2 fixation was thought to occur in soil or possibly in the rhizosphere. Nitrogenase activity associated with washed roots was first shown with Paspalum notatum (15). Still, the observation of live bacteria filling root cells (16) was received with scepticism. Now the infection of roots by Azospirillum has been confirmed in monoxenic cultures (37; 31; 29) and its presence in the steele of field grown maize and other C_4 grasses demonstrated (31; 24; 25). Characteristic root hair branching with certain *Azospirillum* strains and not with others and increase of root hair numbers seems to indicate hormonal effects (Table 4). Most remarkable seems the localization in protoxylem vessels which cab be completely filled over long distances (17; 20) indicating longitudinal distribution. This is confirmed by the appearance of *Azospirillum* in the xylem of stems of maize (25) and wheat (23).

Table 4. Effect of *A. brasilense* on root hair deformations in monoxenic wheat (from 32).

	Control	Sp 7	Sp 107 st[a]	Sp 245[a]
Total nº of root hairs/cm	720	2613	4457	2489
Nº of biforkated root hairs	22	85	157	199

[a] Strains isolated from surface sterilized wheat roots

Carbon substrates from the plant

Malic acid is an important constituent of maize xylem sap (9) and other organic acids may be important. Trans-aconitate was found predominant in aereal parts (10) and roots (D.K. Stumpf person. comm.) of maize. Thirteen out of 17 *A. brasilense* and 13 of 14 *A. lipoferum* strains and 8 out of 9 *A. amazonense* strains use trans-aconitate as sole carbon source (Boddey & Döbereiner 1983 in preparation) as well as 2 *Azotobacter paspali* and one *Derxia gummosa* strain tested. Several other N_2 fixing bacteria, *Beijerinckia indica*, *B. comargense*, *Xantobacter autotrophicum*, *X. flavum*, *Klebsiella pneumoniae*, *Rhizobium* spp. and *E. coli* did not use it.

The hypothesis of the xylem being the principal site of associative N_2 fixation is very attractive not only due to the availability of a constant flow of organic acids and the possibility of fixed N being directly transported ro the shoots by the transpiration stream, but also because low pO_2 access is more probable to occurr there were organisms packed into tubes escavenge rapidly the limited O_2 which diffuses through the endoderm and root cortex tissues.

Plant-Bacteria interactions

Host plant specificity as observed by (2) and (34) has not been confirmed by other laboratories (28) although there are several reports of strains isolated from the same host being more effective inoculants than others (35; 41). The preferential establishment of *A. lipoferum* within inoculated maize roots in the field, even when an inoculant containing *A. brasilense* Sp 7 and Cd was applied which established on the root surface is another confirmation of host plant specificity (Table 5). Strain Sp 242 on the other hand does establish within roots. This strain showed most repeatable inoculation effects (14) in maize, in our hands.

In a large number of field experiments in Israel large and statistically significant *Azospirillum* inoculation effects have been observed with strains Sp 7 and Cd (22). Recent results from our laboratory (20; 3) confirm field inoculation effects in the order of 40 kg N/ha with homologous but not with heterologous

Table 5. Establishment of *Azospirillum* inoculant strains in field grown maize (V.L.D. Baldani, J.I. Baldani & J. Döbereiner, unpublished)

Inoculant strain	Nº of isolates	A. brasilense %	A. lipoferum %
Washed roots			
Control	38	32 nir⁻	68 nir⁻
Sp 7 + Cd	34	76 nir⁺	24 nir⁻
Sp 242[a]	18	66 nir⁻	34 nir⁻
Surface sterilized roots[b]			
Control	20	15 nir⁻	85 nir⁻
Sp 7 + Cd	18	28 nir⁺	72 nir⁻
Sp 242[a]	10	80 nir⁻	20 nir⁻

[a] See table 3
[b] 30 min. in choramine t (1%)

strains. The correlation of numbers of A. *brasilense* in surface sterilized (15 min.) but not in washed roots with the total N incorporation in wheat give support to the main role of internally located organisms even if they are in lower numbers (3).

Whether *Azospirillum* inoculation effects on N incorporation of plants can always be attributed to N_2 fixation seems now doubtfull. Okon (28) suggested a kind of a sponge effect due to growth substances which enhance nutrient uptake in general. The possible role of enhanced NO_3^- reduction and incorporation emmerges from the data on Table 6 where the plants inoculated with *Azospirillum* (especially with the homologous strains) assimilated more $^{15}NO_3$ and instead of the expected isotope dillution in fact a slight increase in ^{15}N% excess is observed.

Studies with nitrate reductase negative mutants of A. *brasilense* give support to the role of the bacterial nitrate reductase in plant NO_3^- uptake (Table 7). The plants inoculated with NR^- mutants reduce more NO_3^- in their leaves than plants inoculated with the NR^+ parent strain.

Quantification of N_2 Fixation

Appart of the above discussed effects of *Azospirillum* on plants there is now unquestionable direct evidence of N_2 fixation in association with grasses and cereals. The initial observations with the C_2H_2 reduction method have now been confirmed by $^{15}N_2$ incorporation (12; 21; 19) by ^{15}N dillution (6; 7) and by N balance studies (1). These studies show values between 10-30% of the total N incorporation being due to biological fixation, in natural systems which have not been inoculated. Better understanding of these systems should permit

Table 6. Nitrogen accumulation and N incorporation from $^{15}NO_3^-$ in grains of field grown wheat inoculated with various *Azospirillum brasilense* strains (R.M. Boddey, V.L.D. Baldani & J. Döbereiner in preparation)

Inoculum	dry wt. g/cylind.	% N	Total N mg/cylind.	% ^{15}N excess	^{15}N recovered mg/cylind.
Sp 107 st[a]	42,3	2,83	1195	0,190	2,30
Sp 245[a]	46,7	2,72	1271	0,171	2,20
Sp 7	45,2	2,81	1276	0,159	2,00
Control	39,3	2,21	866	0,156	1,33
LSD (Tuckey)	N.S.	0,33	318	N.S.	0,67

[a] Isolated from surface sterilized wheat roots

Table 7. The role of A. *brasilense* nitrate reductase in NO_3^- assimilation by monoxenic wheat plants (M.C.B. Ferreira & J. Döbereiner 1983, unpublished)

Inoculant strain	Nitrogenase act. nmol C_2H_4/h/cult.		Nitrate reductase act. μmol NO_2^-/mg leaf tissue
	2 ppm N	20 ppm N	20 ppm N
Sp 7 nir$^+$	9,9	0,0	0,27
Sp 242^a nir$^-$	4,7	0,0	0,19
Sp 107 st^b nir$^-$	12,0	0,0	0,12
Sp 245^b nir$^-$	6,2	0,0	0,15
Sp 245 nr$^-_3$ ^{c}	4,3	0,0	0,50
Sp 245 nr$^-_{15}$	4,8	6,9	0,60
Sp 245 nr$^-_{16}$	2,0	4,4	0,33
Sp 245 nr$^-_{18}$	3,2	0,7	0,32
Sp 245 nr$^-_{22}$	4,8	0,2	0,32
Plant only	0,0	0,0	0,22

[a] See table 3
[b] See table 5
[c] NR$^-$ mutants obtained as described by Magalhães *et al.* (27)

increases of these amounts which reach economic importance.

It is interesting to observe how many of the rather preliminar observations presented at the 1974 N_2 fixation Conference have now been well established. Still we are only beginning to understand the mechanism of the *Azospirillum* associations. No good data about the infection process are available and nothing is known about the mechanism of the associations. Most urgently is a reliable method needed which permits screening of large numbers of plant genotypes. *Azospirillum* physiology studies and especially genetics which may help to make this organism more efficient in several ways are an open field. Certainly some progress will result from this Workshop.

References

1. App, A.A., Watanabe, I., Alexander, M., Ventura, W., Daez, C., Santiago, T. and De Datta, S.K. 1980. Soil Sci. 130. 283-289.
2. Baldani, V.L.D. and Döbereiner, J. 1980. Soil Biol. Biochem. 12, 433-439.
3. Baldani, V.L.D., Baldani, J.I. and Döbereiner, J. 1983. Can. J. Microbiol. (in press).
4. Becking, J.H. 1963. J. Microbiol and Serol. 29, 326.
5. Beijerinck, M.W. 1925. Centralb. für Bakt. etc. II. Abt. Bd. 63, nº 18/22, 353-359.
6. Boddey, R.M., Chalk, P.M., Victoria, R. and Matsui, E. 1983. Soil Biol. Biochem. 15, 25-32.
7. Boddey, R.M., Chalk, P.M., Victoria, R.L., Matsui, E. and Döbereiner, J. 1983. Can. J. Microbiol. (in press).
8. Boddey, R.M. and Döbereiner, J. 1982. In: Non-Symbiotic Nitrogen Fixation and Organic Matter in the Tropics, New Delhi, India, 28-47.
9. Butz, R.G. and Long, R.C. 1979. Plant Physiol. 64, 684-689.
10. Clark, R.B. 1968. Crop. Sci. 8, 165-167.
11. Day, J.M. and Döbereiner, J. 1976. Soil Biol. Biochem., 8, 45-50.
12. De-Polli, H., Matsui, E., Döbereiner, J. and Salati, E. 1977. Soil Biol. Biochem. 9, 119-123.
13. Döbereiner, J. 1978. In: Environmental Role of Nitrogen-Fixing Blue-green algae and asymbiotic bacteria (U. Granhall, ed.), Ecol. Bull. 26, 343-351.
14. Döbereiner, J. and Baldani, J.I. 1982. Ciência e Cultura 34, 869-881.
15. Döbereiner, J., Day, J.M. and Dart, P.J. 1972. J. Gen. Microbiol. 71, 103-116.
16. Döbereiner, J. and Day, J.M. 1976. In: Proceedings fo the 1st International Symposium on N_2 Fixation (W.E. Newton and C.J. Nyman, eds.), Washington State University Press, Pullman, 518-537.

17. Döbereiner, J. and De-Polli, H. 1981. In: The Soil/Root System in Relation to Brazilian Agriculture (R.S. Russell, K. Igue and Y.R. Mehta, eds.), Fundação Instituto Agronômico do Paraná, Londrina, Paraná, 175-198.
18. Döbereiner, J., Marriel, I.E. and Nery, M. 1976. Can. J. Microbiol. 22, 1464-1473.
19. Eskew, D.L., Eaglesham, R.J. and App, A.A. 1981. Plant Physiol. 68, 48-52.
20. Freitas, J.L.M. de, Rocha, R.E.M. da, Pereira, P.A.A. and Döbereiner, J. 1982. Pesq. agropec. bras. 17, 1423-1432.
21. Ito, O., Cabrera, D. and Watanabel, I. 1980. Appl. Environ. Microbiol. 39, 554-558.
22. Kapulnik, Y., Sarig, S., Nur, I., Okon, Y., Kigel, J. and Henis, Y. 1981. Expl. Agric. 17, 179-187.
23. Kavimandan, S.K., Lakshmi-Kumari, M. and Subba-Rao, N.S. 1978. Proc. Indian Acad. Sci. Sect. B, 87, 299-302.
24. Lakshmi, V., Rao, A.S., Vijayalaksmi, K., Lakshmi-Kumari, M., Tilak, K.V.B.R. and Subba-Rao, N.S. 1977. Proc. Indian Acad. Sci. 86, 397-404.
25. Magalhães, F.M.M., Patriquin, D. and Döbereiner, J. 1979. Rev. Brasil. Biol. 39, 587-596.
26. Magalhães, F.M.M., Baldani, J.I. and Döbereiner, J. 1983. An. Acad. Brasil. Ciê. (in press).
27. Magalhães, L.M.S., Neyra, C.A. and Döbereiner, J. 1978. Arch. Microbiol. 117, 247-252.
28. Okon, Y. 1982. Israel J. Bot. 31, 214-220.
29. Okon, Y., Albrecht, S.L. and Burris, R.H. 1977. Appl. Environ. Microbiol. 33, 85-88.
30. Patriquin, D.G. 1981. In: Advances in Agricultural Microbiology (N.S. Subba Rao, ed.) Publ. M/S Oxford and IBH New Delhi.
31. Patriquin, D.G. and Döbereiner, J. 1978. Can. J. Microbiol. 24, 734-742.
32. Patriquin, D.G., Döbereiner, J. and Jain, D.K. 1983. Can. J. Microbiol. (in press)
33. Reynders, L. and Vlassak, K. 1982. Plant Soil 66, 217-223.

34. Rocha, R.E.M. da, Baldani, J.I. and Döbereiner, J. 1981. *In*: Associative N_2 Fixation (P.B. Vose and A.P. Ruschel, eds.) CRC Press, Boca Raton, Florida, Vol. II, 67-69.
35. Subba-Rao, N.S. 1981. *In*: Associative N_2 Fixation (P.B. Vose and A.P. Ruschel, eds.) CRC Press, Boca Raton, Florida, Vol. I, 137-143.
36. Tarrand, J.J., Krieg, N.R. and Döbereiner, J. 1978. Can. J. Microbiol. **24**, 967-980.
37. Umali-Garcia, M., Hubbell, D.H., Gaskins, M.H. and Dazzo, F.B. 1980. Appl. Environ. Microbiol. **39**, 219-226.
38. Van Berkum. P. and Bohlool, B.B. 1980. Microbiol. Rev. **44**, 491-517.
39. Vargas, M.A.T. 1977. Tese M.S., Univ. Wisconsin, 1-69.
40. Volpon, A.G.T., De-Polli, H. and Döbereiner, J. 1981. Arch. Microbiol. **128**, 371-375.
41. Vlassak, K. and Reynders, L. 1978. *In*: Isotopes in Biological Dinitrogen Fixation-Proceedings of an Advisory Group Meeting, IAEA-Vienna, 71-87.

GENETIC ANALYSIS IN AZOSPIRILLUM

M.BAZZICALUPO AND E.GALLORI

Istituto di Anatomia Comparata,Biologia Generale e Genetica,via Romana 19
University of Florence,Florence,Italy

Introduction
Azospirillum is a bacterium able to fix nitrogen in association with roots of several grasses,stimulating,for this reason,considerable attention of scientists all over the world.
  Genetic studies in *Azospirillum* were developed only in the past few years. Works,carried out mainly in the laboratory of C.Elmerich, demonstrated the possibility to transfer R type plasmids between different strains of *Azospirillum* (1,2),the ability of R plasmids to mobilize chromosomal fragments and the occurrence of linkage between markers(2).
  Transduction has not been reported,although an *Azospirillum* bacteriophage has been isolated and characterized(3).
  Early report on the occurrence of transformation with chromosomal DNA has not been confirmed later on (4).
  Recentely,molecular biology techniques allowed to isolate and study *nif* genes in *Azospirillum*(5).

A.brasilense mutants
  All the mutants used in the works described,derived from parental strains Sp6 (6) and Sp7 (7).Auxotrophic mutants were obtained by Nitroso= guanidine or EMS mutagenesis followed by Cycloserine enrichment,while strains resistant to antibiotics were isolated as spontaneous mutants. The reversion frequencies of the mutation isolated range ,according to the strains,from 10^{-5} to 10^{-9} ;in the conjugation experiments only mutations with reversion frequency lower than 10^{-8} were used. Mutants derived from strains Sp6 and Sp7 behave in a quite similar way in respect to conjugation with plasmid R68-45,and markers from Sp6 express apparently well in strains derived from Sp7,and *vice versa*.

Transfer of plasmid R68-45
  Plasmid R68-45 (*km,tc,cb,tra,incP*) was a gift of B.Holloway,and was transferred by conjugation from *P.aeruginosa* to *A.brasilense* Sp6. Matings were carried out in solid medium(1),although also a low level of transfer in liquid unshaken cultures was found.
  In order to study the kinetic of plasmid transfer,two types of experi= ments were carried out.In the first experiments,cells of the recipient

strain were plated on "Millipore" filters; at time 0 an equal number of donor cells was added to the filets. At intervals the filters were removed from the plates and dipped in a large volume of saline solution, in order to dilute the cells and avoid remating. Appropiate dilution were then plated on selective medium (exp.1,2 and 3,Table 1). In the second type of experiments recipient strain was resistant to nalidixic acid, an inhibitor of DNA synthesis; matings were carried out as before except that at intervals nalidixic acid, to a final concentration of 0.5 mg/ml, was added to the plates, to stop mating by the block of DNA synthesis in the donor cells(8). Cells were collected and plated on selective medium as before (exp.4 and 5,Table 1).

Table 1: Frequency of R68-45 exconjugants in interrupted mating

Time	Frequency of exconjugants				
	exp.1	exp.2	exp.3	exp.4	exp.5
10'	6×10^{-5}	1×10^{-4}	6×10^{-5}	1.3×10^{-7}	6.2×10^{-7}
30'	2×10^{-4}	5×10^{-4}	2×10^{-4}	3.5×10^{-7}	2.4×10^{-6}
45'	-	-	-	2.1×10^{-6}	3×10^{-6}
60'	7×10^{-4}	1×10^{-3}	3×10^{-4}	4×10^{-6}	1.1×10^{-5}
120'	-	-	4.5×10^{-4}	-	-
240'	1.5×10^{-3}	3×10^{-3}	-	-	-

Exconjugants at 0 time were always lower than 10^{-7}, and their frequencies were subtracted.

Data obtained indicate that the transfer of the plasmid starts almost immediately after the contact of the cells and the frequency of exconjugants raises quickly during the first hour; from this time to 10-16 hours the frequency of exconjugants increases up to about 10^{-2} (data not shown in Table 1). Data reported indicate that the frequency of exconjugants differs according to the strain used and was extremely low in experiments 4 and 5 in which the recipient strain was resistant to nalidixic acid.

Transfer of the chromosome

The transfer of the chromosomal markers was carried out as described for the transfer of the plasmid, except that the mating plates were incubated for 10-12 hours.

The frequency of recombination obtained was quite low ranging from 10^{-7}

to 10^{-6}, whatever the marker selected, corroborating the hypothesis of the multiple origin of transfer (2). Plasmid phenotype was found in about 60-70% of the recombinants; however could be initially present in all the recombinants and could be lost during selection and isolation which were carried out in media without antibiotics.

In Table 2 are summarized the results of several conjugation in which the transfer of two or three markers was analyzed. Linkages between selected and unselected markers were calculated by the frequency of the transfer of unselected markers(cotransfer); map distances were expressed as 100-cotransfer. Table 2 also shows that some markers seem not linked.

Table 2: Map distances between linked markers.

markers	distance	markers	distance	unlinked markers+
cys-arg	77	cys-met	88	his3-cys
leu-arg	70	trpE-cys	75	his3-arg
trpE-arg	87	aroA-cys	59	leu-cys
aroA-arg	83	his1-arg	86	his4-arg
his1-cys	84	his1-ade1	43	his4-cys
his1-trpA1	62	his1-trpE	90	his4-aroB
aroB-arg	38	aroB-cys	82	his2-leu
trpA2-arg	63	trpA2-cys	91	
ser-ade2	20	ser-his2	13	+= <1%cotransfer
trpA3-arg	50	trpA3-cys	83	

Numbers following genetic symbols refer to independent mutations

Crosses, in which the simultaneous transfer of two or three markers was analyzed, could be used for ordering the markers on the chromosome, as illustrated by results summarized in Table 3.

Table 3: Order of markers derived from three and four factor crosses

his1---ade1---arg	trpA1---his1---ade1---arg
cys---his1---ade1---arg	his1---ade1---cys
cys---arg---trpE	aroA---cys---arg
cys---arg---aroB	cys---trpA2---arg
cys---trpE---arg	

Data obtained so far are not sufficient to draw conclusions on the organi= zation of the chromosome of *Azospirillum*,but clearly indicate the possibility to construct,in the near future,a genetic map.In some instances data on the order of the markers seem contradictory in different experiments,that could be explained considering that the chromosome of *Azospirillum* is circular and the transfer occurs in both directions.

Further data on the mapping of the genes of *A.brasilense* could be obtained by interrupted mating experiments.Unlukily,at the present,the frequency of recombination is too low to allow these experiments;in particular recombinants in the first two hours after the start of the mating ,were never observed.Selection of mutant plasmids with enhanched chromosome donor ability,would be very useful in obtaining the genetic map of *Azospirillum*.

Conjugation experiments could also be utilized to distinguisch markers with the same phenotype. Table 4 reports results of crosses between strains carrying independent *trp* mutations.Data demonstrate that *trpA1* produces Trp^+ recombinants in crosses with *trpE* and *trpC1,trpC2* while recombinants in crosses with other *trpA* markers were not found.

Table 4: Analysis of *trp* loci

Donor	Recipient (relevant genotype)	Trp^+ recombinants (frequency)
trpA1	*trpE*	3×10^{-7}
"	*trpC1*	2×10^{-7}
"	*trpC2*	4×10^{-7}
"	*trpA3*	$< 10^{-8}$
"	*trpA4*	$< 10^{-8}$
"	*trpA5*	$< 10^{-8}$

trpA,*trpC* and *trpE* indicate requirements of: triptophan ,indole or trip= tophan,anthranilate or indole or triptophan,respectively.

References

1. Polsinelli,M.,Baldanzi,E.,Bazzicalupo,M.and Gallori,E. (1980) Molec.G.Genet. 178:709-711.

2. Franche,C.,Canelo,E., Gauthier,D.and Elmerich,C.(1981) FEMS Microbiol.Lett. 10:199-202.

3. Elmerich,C.,Quiviger,B.,Rosemberg,C.,Franche,C.,Laurent,P. and Dobereiner,J. (1982) Virology 122:29-37.

4. Mishra,A.K.,Roy,P. and Bhattacharya,S. (1979) J.Bacteriol.137:1425-1427.

5. Quiviger,B., Franche,C.,Lutfalla,G.,Rice,D.,Haselkorn,R. and Elmerich,C. (1982) Biochimie 64:495-503.

6. Bani,D.,Barberio,C.,Bazzicalupo,M.,Favilli,F.,Gallori,E. and Plosinelli,M. (1980) J.Gen.Microbiol. 119:239-244.

7. Tarrand,J.,Krieg,N. and Dobereiner,J. (1978) Can.J.Microbiol.24:967-980.

8. Haas,D. and Holloway,B.(1976) Mol.Gen.Genet.144:243-251.

RECENT DEVELOPMENTS IN THE GENETICS OF NITROGEN FIXATION IN *AZOSPIRILLUM*

S.K. NAIR, P. JARA, B. QUIVIGER and C. ELMERICH

Unité de Physiologie Cellulaire, Département de Biochimie et Génétique Moléculaire, Institut Pasteur, 28 Rue du Dr Roux 75724 Paris, Cedex 15, France

Introduction

In *Klebsiella pneumoniae* the structural genes for the nitrogenase complex, *nifHDK*, belong to the same operon and are carried by a single 6.2 kb EcoRI fragment, which was cloned in plasmid pSA30 (1). Using this plasmid as a probe, homology was found with total DNA of a large number of diazotrophs (2,3,4), including *Azospirillum brasilense* and *A. lipoferum* (5,6). In order to clone *Azospirillum* DNA and to initiate genetic analysis of nitrogen fixation in this bacterium, we looked for homology with a series of *K. pneumoniae nif* probes covering the entire *nif* cluster. In addition Nif⁻ mutants were analyzed by complementation after construction of *nif* partial diploids.

Homology with *K. pneumoniae nif* probes

The fragments of *K. pneumoniae nif* DNA used as probes in this work to study homology with *Azospirillum* DNA are schematized in Figure 1. Strong homology was detected with pSA30 that carries the *nifHDK* structural genes for the nitrogenase components, as previously reported (5,6). Weak homology was detected with plasmid pPC936 and with plasmid pMC71A that both carries the *nifA* gene coding for a *nif* specific activator (11). Figure 2 shows the results of hybridization with the pMC71A plasmid. The size of the fragments detected in *A. brasilense* and *A. lipoferum* is different. This suggest a different gene organization in the two species, as already observed with the *nifHDK* genes (6). With plasmid pPC880, that carries the *nifJ* gene, homology was detected only with *A. lipoferum* Br17. The size of the fragments was of 10.2 kb for BamHl; 9.4 kb for BglII and HindIII; and 6.6 kb for EcoRI. No homology was found with plasmids pPC937 and pGR113 in the case of both strains.

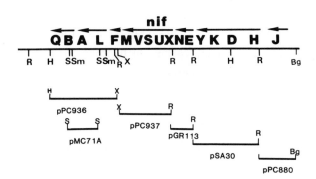

Figure 1 : *K. pneumoniae nif* probes used in the hybridization studies. The *nif* cluster organization is drawn according to references 1,7-10. Plasmids pPC937 and pPC880 were constructed for this work, other plasmids are described in references 1 (pSA30, pGR113) ; 10 (pPC936) ; 11 (pMC71A). Restriction sites : Bg : *Bgl*II ; H : *Hin*dIII ; R : *Eco*RI ; S : *Sal*I ; Sm : *Sma*I.

Cloning of *nif* DNA

A phage library containing *Eco*RI inserts of strain 7000 DNA was constructed in the λgt7-*ara6* vector (6). This vector has a cloning capacity for *Eco*RI fragments from 2 to 14 kb by substitution of the *ara6* fragment.

Cloning of a *nifHDK* cluster

Using plasmid pSA30 as a probe, a recombinant phage carrying the 6.7 kb *Eco*RI fragment, designated AbRI, was isolated by *in situ* plaque hybridization (14). Heteroduplex DNA experiments showed extensive sequence homology between the AbRI fragment and the *K. pneumoniae nifHDK* probe suggesting the existence in strain 7000 of a *nifHDK* cluster (6). The physical map of the AbRI fragment is shown in Figure 3. The direction of transcription as well as the gene approximate location were deduced from the heteroduplex experiments (6).

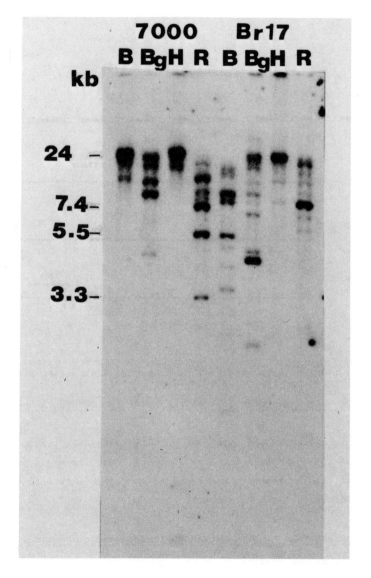

Figure 2 : Hybridization of total DNA of *A. brasilense* 7000 (ATCC29145) and *A. lipoferum* Br17 (ATCC29709) with the *nifA* probe from *K. pneumoniae*. Total DNA was purified according to (6), hydrolyzed with restriction enzymes B (*Bam*HI), Bg, H, R. The probe was the *Sal*I fragment of plasmid pMC71 purified from an agarose gel by electroelution and labelled by nick translation (12) with $\alpha^{32}P$ deoxyadenosine triphosphate. Hybridization was performed under non stringent conditions according to Southern (13).

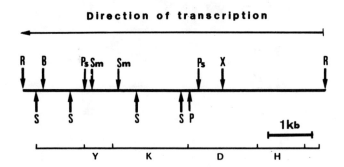

Figure 3 : Organization of the AbRI fragment. Restriction sites as in Figure 1 and 2, P : PvuII, Ps : PstI, X : XhoI.

Cloning a DNA fragment that shares homology with the K. pneumoniae nifA gene

The phage library was also used to perform *in situ* plaque hybridization with plasmid pMC71A as a probe. A phage that carries a 7.4 kb EcoRI fragment was purified. The fragment was found to share homology with the nifA probe by hybridization (data not shown).

Characterization of a Nif⁻ mutant by genetic and biochemical complementation

- Isolation of mutants

After EMS mutagenesis of strain 7000, we isolated mutants with Nif⁻ phenotype (15). However, only one of them, strain 7571, was characterized as impaired in a *nif* gene.

- Genetic complementation

Genetic complementation was performed by construction of *nif* partial diploids. The AbRI fragment was cloned at the unique EcoRI site of plasmid pRK290 to yield pAB35 (15). Plasmid pRK290 is a broad host range vector which can be transferred by conjugation into Gram negative bacteria using another plasmid, pRK2013, as mobilizing agent (16). Plasmid pAB35, when introduced in strain 7571, restored nitrogen fixation as tested in whole cells assay and in crude extracts (Table 1). A derivative of pAB35, designated pAB36, which was deleted of the 2.3 kb PstI fragment that likely carries nifK and part of nifD, but still contained nifH (see Figure 2) was constructed (15). This plasmid did not restore nitrogen fixation in strain 7571.

Table 1 : Biochemical and genetic complementation of the Nif⁻ mutant 7571.

Strain	Nitrogenase activity (U/mg protein)			
	Whole cells	Crude extracts no addition	+ Kp1	+ Kp2
7000	70	7.8	7.6	7.8
7571	< 0.01	< 0.01	1.0	< 0.01
7571(pAB35)	45	6.6	ND	ND
7571(pAB36)	< 0.01	< 0.01	0.3	< 0.01

Bacteria from 300 ml minimal medium plus NH_4^+ (17) were centrifuged and inoculated to 1,5 l fermentor containing 1 l of N free medium at $OD_{570\ nm}$: 0.4. The culture was incubated under an N_2-O_2 atmosphere (99-1%), with stirring, at 30°C. After 6 h derepression, nitrogenase was assayed on aliquots and the cells were collected anaerobically. Crude extracts were obtained by passage through a French pressure cell and nitrogenase was assayed according to Okon et al (18). The amount of protein in the assay was about 2 mg. Pure nitrogenase components added for complementation, Kp1 (48 µg) and Kp2 (20 µg), were prepared by J. Houmard. One unit of activity corresponds to one nmole C_2H_2/min/mg protein. ND : not determined.

- Biochemical complementation

Biochemical complementation was tested on crude extracts of cells incubated under conditions of nitrogenase derepression, by addition of pure nitrogenase components of K. pneumoniae. As shown on Table 1, nitrogenase activity was restored by addition of pure Kp1 but not of pure Kp2.

- Immunoprecipitation of nitrogenase polypeptides

To determine whether or not the nitrogenase polypeptides were synthesized in strain 7571, antisera against both components of K. pneumoniae were used in immunoprecipitation experiments. The molecular weight of the polypeptides detected was compared to those obtained after labelling minicells containing plasmid pSA30 (see Figure 4). With anti-Kp2 antiserum, two polypeptides of 36,000 and 39,000 daltons were precipitated from the wild type and the mutant. The two polypeptides are likely the components of nitrogenase protein 2. The presence of two bands for component 2 was also reported for some photosynthetic bacteria (19). With anti-Kp1 antiserum, the intensity of the polypeptides bands detected varied from one experiment to another. Some of the polypeptides, in particular those of 92,000, 80,000, 56,000 and 54,000 daltons were in some experiments precipitated from ammonia grown cultures and were also detected with anti-Kp2 antiserum. It was assumed that these polypeptides were not specific for nitrogen fixation. However, the 64,000 and 60,000 dalton bands correspond likely to the nitrogenase component 1 polypeptides.

Discussion

In order to obtain more information about the genetic organization of *Azospirillum nif* genes, we tried to develop a methodology based on *nif* DNA cloning and on the use of cloned DNA to perform genetic complementation of *nif* mutants.

Hybridization experiments with *K. pneumoniae nif* probes revealed the existence of a *nifHDK* cluster in *A. brasilense* 7000 (6). We report here homology with the *nifA* activator gene. In earlier experiments, homology between *K. pneumoniae nif* DNA and total DNA of a series of diazotrophs was found to be limited to *nifH* and *D* (4). It appears now that homology can be extended to other *K. pneumoniae nif* genes by using specific probes of smaller size. Homology was observed with *nifK* in several diazotrophs (see i.e. 7, 21-24) *nifV* and *S* in cyanobacteria (21) and with *nifA* of *Rhizobium leguminosarum* (Johnston, First Insternational Symposium on Bacteria - Plant Interaction, Bielefeld FRG, 1982). In addition, homology was also recently detected with total DNA of a tropical *Rhizobium* and *K. pneumoniae nifNE* and *nifJ* probes (F. Norel, this laboratory unpublished).

It is still too early to know, in the case of *A. brasilense* 7000, if the 7.4 kb *Eco*RI fragment which was cloned and which shared homology with *nifA*,

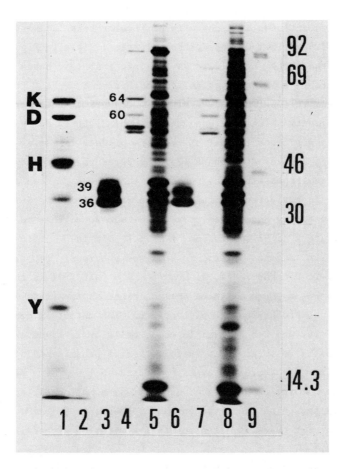

Figure 4 : Immunoprecipitation of nitrogenase components of strains 7000 and 7571. Cells incubated under conditions of nitrogenase derepression were labelled with 20 µCi ^{35}S methionine. Nitrogenase components were immunoprecipitated using K. pneumoniae anti-Kp1 and anti-Kp2 antisera (prepared by J. Houmard) in the presence of Staphylococcus aureus cells (20). Samples were loaded on 10% polyacrylamide gels and the gels were impregnated for fluorography and exposed to X Ray films. 1 : minicells of E. coli containing pSA30 ; 2-5 strain 7000 ; 6-8 strain 7571 ; 9 : MW markers by decreasing size : phosphorylase b, bovine serum albumin, ovalbumin, carbonic anhydrase and lysozyme. 2 : control of precipitation without antiserum ; 3 and 6 : immunoprecipitation by anti-Kp2 antiserum ; 4 and 7 : immunoprecipitation by anti-Kp1 antiserum ; 5 and 8 : not immunoprecipitated samples.

carries the equivalent of a *nifA* gene. In particular three *Eco*RI fragments of respectively 3.5 kb, 5.1 kb and 7.4 kb, clearly visible in Figure 2, hybridized with the probe. Even if it is possible that some bands were due to partial digest, it remains to establish whether or not these fragments correspond to gene reiteration. Moreover, further work is necessary to prove that the DNA fragments which were detected by hybridization carry gene(s) whose function is analogous to that of the *nifA* product (see i.e. 10,11). In a recent paper, Kennedy and Robson (25) suggested that *nifA* may be ubiquitous among diazotrophs since they observed that the *nifA* gene of *K. pneumoniae* could activate *nif* transcription in *Azotobacter*. Similar results were reported by Sundaresan et al. (26) who found that *K. pneumoniae* *nifA* could activate a *nifH::lac* fusion of *R. meleloli* in *E. coli*.

Isolation of Nif⁻ mutants of *Azospirillum* was previously reported (15, see also this book). To our knowledge, *A. brasilense* strain 7571 is the only mutant of *Azospirillum* which has been characterized as impaired in a *nif* gene by genetic and biochemical complementation. The mutant is devoid of nitrogenase activity whether assayed in whole cells or in crude extracts. Nitrogen fixation is restored after introduction of plasmid pAB35 that carries the entire *nifHDK* cluster of *A. brasilense* but not by a plasmid deleted of *nifK* and part of *nifD* gene. Biochemical complementation by Kp1 and immunoprecipitation of the nitrogenase components by *K. pneumoniae* antisera strongly suggest that the mutant carries a mutation either in the *nifD* or the *nifK* gene.

For further developments of genetics of nitrogen fixation it will be necessary to obtain a large number of *nif* mutants and to characterize them biochemically. Concerning the procedure of mutants isolation, the use of cloned *nif* DNA to perform Tn5 site directed mutagenesis, according to the methodology described in *Rhizobium* (24,27, see also this book), should allow to construct rapidly a large collection of mutants. Biochemical identification of the *nif* mutants appear to be more difficult. First, the nitrogenase complex of *Azospirillum* is not well known, no specific antisera are available, and up to now specific activities detected in crude extracts are roughly 10% of those detected *in vivo*. Second, regulation of nitrogenase activity, which is submitted to *in vivo* ammonia switch off (Jara and Elmerich in preparation) is still to be understood. Third, even though we used immunoprecipitation with anti-Kp1 and anti-Kp2 immusera, the identification of the nitrogenase complex polypeptides of *A. lipoferum* and *A. brasilense* need further confirmation and no other polypeptides have been identified yet.

Acknowledgments

This work was supported by funds from the University Paris VII and by a research contract from Elf Bio-Recherche, Entreprise Minière et Chimique, Rhône-Poulenc Recherches and CDF Chimie.

References

1. Riedel, G.E., Ausubel, F.M. and Cannon, F.C. 1979. Proc. Natl. Acad. Sci. U.S.A. 76, 2866-2870.
2. Nuti, M.P., Lepidi, A.A., Prakash, R.A. and Cannon, F.C. 1979. Nature (London) 282, 533-535.
3. Mazur, B.J., Rice, D., and Haselkorn, R. 1980. Proc. Natl. Acad. Sci. U.S.A. 77, 186-190.
4. Ruvkun, G.B. and Ausubel, F.M. 1980. Proc. Natl. Acad. Sci. U.S.A. 77, 191-195.
5. Elmerich, C. and Franche, C. 1982. In "*Azospirillum* Genetics, Physiology, Ecology" Klingmüller, W., ed., Birkhäuser Basel EXS42, pp. 9-17.
6. Quiviger, B., Franche, C., Lutfalla, G., Rice, D., Haselkorn, R. and Elmerich, C. 1982. Biochimie 64, 495-502.
7. MacNeil, T., MacNeil, D., Roberts, D., Supiano, M.A. and Brill, W.J. 1978. J. Bacteriol. 136, 821-829.
8. Merrick, M., Filser, M., Dixon, R., Elmerich, C., Sibold, L. and Houmard, J. 1980. J. Gen. Microbiol. 117, 509-520.
9. Pühler, A. and Klipp, W. 1980. In "Biology of Inorganic Nitrogen and Sulphur Compounds" Bothe A. and Trebst, A. eds. Springer, Berlin Heidelberg New York pp. 276-286.
10. Sibold, L., Quiviger, B., Charpin, N., Paquelin, A. and Elmerich, C. 1983. Biochimie 65, 53-63.
11. Buchanan-Wollaston, V., Cannon, M.C. and Cannon, F.C. 1981. Nature (London) 294, 776-778.

12. Rigby, P.W.J., Dieckmann, M., Rhodes, C. and Berg, P., 1977. J. Mol. Biol. 113, 237-251.
13. Southern, E.M. 1975. J. Mol. Biol. 98, 503-507.
14. Benton, W. and Davis, R., 1977. Science 196, 180-181.
15. Jara, P., Quiviger, B., Laurent, P. and Elmerich, C. 1983. Can. J. Microbiol. 29, in press.
16. Ditta, G., Stanfield, S., Corbin, D., and Helinski, D. 1980. Proc. Natl. Acad. Sci. U.S.A. 77, 7347-7351.
17. Franche, C. and Elmerich, C. 1981. Annal Microbiol. 132A, 3-19.
18. Okon, Y., Houchins, J.P., Albrecht, S.L. and Burris, R.H. 1977. J. Gen. Microbiol. 98, 87-93.
19. Ludden, P.W., and Burris, R.H. 1978. Biochem. J. 175, 251-259.
20. Kessler, S.W. 1976. J. Immunol. 117, 1482-1496.
21. Rice, D., Mazur, B.J. and Haselkorn, R. 1982. J. Bioch. Chem. 257, 13157-13163.
22. Elmerich, C., Dreyfus, B.L. Reysset, G., and Aubert, J.P. 1982. EMBO J.1 499-503.
23. Fuhrmann, M. and Hennecke, H. 1982. Mol. Gen. Genet. 187, 419-425.
24. Ruvkun, G.B., Sundaresan, V. and Ausubel, F.M. 1982. Cell 29, 551-559.
25. Kennedy, C. and Robson, R.L. 1983. Nature London 301, 626-628.
26. Sundaresan, V., Ow, D.W. and Ausubel 1983. Proc. Natl. Acad. Sci. U.S.A. 80, 4030-4034.
27. Simon, R., Priefer, U. and Pühler, A. 1983. In First International Symposium on Bacteria - Plant Interaction - Proceedings, Pühler A. ed Springer, Heidelberg, Berlin, New-York, in press.

MOLECULAR CLONING OF NITROGEN FIXATION GENES FROM AZOSPIRILLUM

W. Wenzel, M. Singh and W. Klingmüller
Lehrstuhl für Genetik, Universität Bayreuth,
Universitätsstraße 30, D-8580 Bayreuth, FRG

Introduction

Nitrogen fixation is catalyzed by nitrogenase which consists of two proteins, and the cistrons coding for the polypeptides are organized in one transcriptional unit, called nif HDKY cluster in K. pneumoniae. The gene nif H codes for the Fe protein subunit, nif D and nif K for the MoFe protein subunits. The nif HDKY operon of K. pneumoniae was cloned into a plasmid pSA30 (1). Since this operon is strongly conserved in evolution, homology of this DNA was found in a number of nitrogen-fixing procaryotes (2). Due to this homology, several groups succeeded in cloning nif HDK homologous DNA from Rhizobium meliloti (3,4), Rhizobium japonicum (5), Anabaena (2) and Azospirillum brasilense strain 7000 (6). The nif HDK region, a 6.7 kb Eco RI fragment from A. brasilense strain 7000 (ATCC 29145), was cloned into a plasmid pAB1 (6).

We used this plasmid for screening our recombinant λ-libraries of Azospirillum DNA for the presence of nif HDK. In K. pneumoniae the nif-genes are tandemly clustered in seven transcriptional units (7). Therefore, the total nif DNA of K. pneumoniae is arranged within a 24 kb fragment of DNA. If a similar organization of nif-genes is there in Azospirillum as well, we should perhaps be able to detect further relevant nif-genes other than the structural nif HDK operon, on inserts larger than 6.7 kb. Therefore, we have chosen phage λ 47.1 (8) for a vector as it allows for the insertion of large Eco RI fragments (8-24 kb). We established a library of recombinant phages from two Azospirillum strains, A. brasilense ATCC 29710 str^r rif^r and

A. lipoferum A 23.

Materials and Methods

Isolation of Azospirillum chromosomal DNA

Bacteria were harvested from a fresh 1 l overnight culture in Luria Broth (pH 7.0), resuspended and washed in 10 ml 50 mM Tris HCl, 25 % sucrose (pH 8.0). Lysis was obtained by adding 50 mg lysozyme and EDTA to a final concentration of 50 mM at $37°$ C for half an hour. After incubation, SDS was added to a final concentration of 2 % and the lysate was incubated for further 30 minutes at $37°$ C. The DNA was purified from this mixture by extraction with phenol and phenol-chloroform (1:1). To avoid sheering, the aquaeous phases were carefully decanted and not pipetted. The DNA was precipitated with ethanol for 3 times. The DNA precipitate was handled with forceps and carefully dissolved in TE, pH 7.6 (10 mM Tris-HCl, 1 mM EDTA) with very gentle shaking. After this precipitation and purification steps, the contaminating RNA was digested with RNase in TE containing 100 μg/ml RNase at $37°$ C for 1 hour. DNA was extracted by two steps of phenol-chloroform treatment and further purified by ethanol precipitation as described.

Preparation of the Azospirillum DNA for molecular cloning

The dried DNA precipitate was dissolved in TE to give a final concentration of 50 μg DNA/ml. The enzymatic conditions for preparing partially Eco RI digested DNA were monitored in a pilot experiment according to Maniatis et al. (9), starting with 1 μg of DNA and 2 U of Eco RI in a twofold serial dilution of the restriction enzyme. At enzyme dilutions of 1:8 and 1:16, the DNA was digested to give fragments of a size of 15-24 kb. According to these results DNA was prepared for cloning at a larger scale of 10 μg for each Azospirillum species.

Preparation of vector DNA

Phage λ 47.1 particles were purified according to Loenen and Brammar (8). The DNA from these phage particles was isolated after a procedure given by Maniatis et al. (9). After complete digestion of the λ 47.1 DNA with Eco RI, samples of 1 µg were loaded onto a 0.4 % preparative agarose gel. The λ arms were purified from this gel by electroelution into dialysis bags (9). The eluate was extracted twice with phenol and once with phenol-chloroform and recovered by two ethanol precipitations. Finally the pellet was rinsed once with 70 % ethanol, dried and re-suspended in 20 µl of TE (pH 7.6).

Ligation of the DNA

Ligation of λ 47.1 arms and Azospirillum DNA fragments was carried out with 3 µg of DNA in a volume of 10 µl of ligation buffer. The molar ratio of arms and inserts was adjusted to 1:1:1. This mixture was incubated at 4° C overnight with 5 units of T4 ligase (Boehringer, Mannheim). Aliquots of the reaction mixture were tested before and after enzyme incubation. λ 47.1 arms and stuffer DNA was tested in a separate ligation for the production of viable phages.

Packaging of the DNA

The ligated DNA was packaged into phage λ particles according to a procedure described by Hohn (10) using E. coli strains BHB2690 and BHB2688 for preparation of the packaging extract. After purification of the recombinant λ particles, the E. coli strain LE392 was infected with these phages and layered onto LB plates. The plates were incubated overnight at 37° C.

Screening for nif HDK containing phages

Two plates with 1000-1200 plaques of each library were screened by in situ hybridization as described by Benton and Davis (11). As a radioactive probe, we used plasmid pAB1 (a gift of Dr. Elmerich), containing a 6.7 kb nif HKD fragment from A. brasilense strain 7000 (6). pAB1 was ^{32}P-labelled by nick-translation after a procedure described by Maniatis et al. (9).

Southern blot analysis of chromosomal Azospirillum DNA

In order to detect fragments of DNA homologous to nif HDK DNA from A. brasilense strain 7000, the DNA of our Azospirillum species was completely digested with Eco RI. After electrophoresis on a 0.7 % agarose gel the DNA was transferred to nitrocellulose paper according to Southern (12). Hybridization of the Southern filters was carried out as described by Maniatis et al. (9).

Results and Discussion

As shown in fig. 1, homology between the pAB1 probe and DNA of A. brasilense strain ATCC 29710 str^r rif^r and A. lipoferum A23 can be detected. In the A. brasilense strain, the homology is restricted to only one Eco RI fragment of about 6.7 kb. In contrast to A. brasilense, the A. lipoferum strain exhibits two distinct fragments of about 9.0 kb and 2.5 kb. Therefore, there could be a further Eco RI site within the nif HDK cluster of A. lipoferum that is lacking in the homologous DNA of the A. brasilense strain. A similar polymorphism in number and size of fragments homologous to nif HDK DNA between different Azospirillum strains was detected by Quiviger et al. (6) as well.

The screening of recombinant phages resulted in 10 plaques hybridizing with pAB1 DNA in the case of our A. lipoferum library and 14 positive plaques in the A. brasilense ATCC 29710 str^r rif^r phage library (fig. 2).

After purification of these plaques, we tested the purified phages for the presence of nif HDK homologous DNA by dot hybridization with radioactively labelled pAB1 DNA (fig. 3). The results clearly indicate that most of the phages picked up from plaques hybridizing to pAB1 contain inserts of homology with nif HDK.

The titer of phages growing up in liquid culture is low and amounts to about 3×10^8 pfu/ml. This may perhaps be due to the nature of the large inserts that may interfere with phage propagation or other cell functions during infection.

We therefore intend to pool the phages which showed a positive signal in the plaque hybridization assay in order to reclone and subclone the phage DNA into amplifiable E. coli plasmids. This would allow for small scale preparation of DNA necessary for the characterization of the DNA inserts by restriction analysis and Southern blot hybridization.

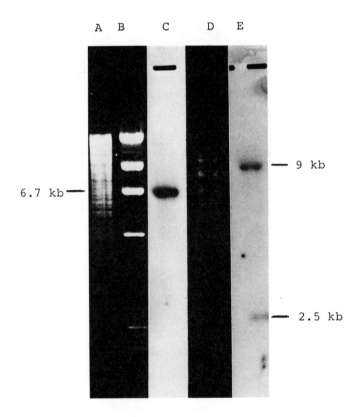

Figure 1: Localization of <u>nif</u> HDK genes on Eco RI fragments of Azospirillum DNA

- A   <u>A. brasilense 29710</u> str^r rif^r DNA digested with Eco RI
- B λ DNA digested with HindIII
- C   Southern autoradiogram of <u>A. brasilense</u> 29710 str^r rif^r DNA digested with Eco RI
- D   <u>A. lipoferum</u> A23 DNA digested with Eco RI
- E   Southern autoradiogram of <u>A. lipoferum</u> A23 DNA digested with Eco RI

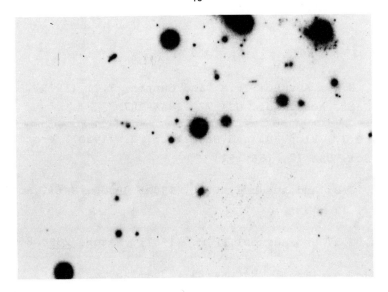

Figure 2: Plaque hybridization of a A. brasilense library with ^{32}P-pAB1 (6.7 kb nif fragment)

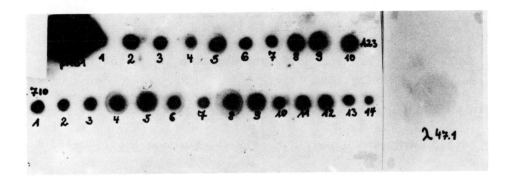

Figure 3: Dot hybridization of recombinant phages with ^{32}P-pAB1 (6.7 kb nif fragment).
The spots A23 and 710 refer to the total DNA from strains A. lipoferum A23 and
A. brasilense ATCC 29710 str^r rif^r, respectively
Titer of phages per dot: 2×10^8 pfu

References

1. Riedel, G.E., Ausubel, F.M. and Cannon, F.C. (1979). Proc. Natl. Acad. Sci. USA **77**, 2866-2870

2. Mazur, B.J., Rice, D. and Haselkorn, R. (1980). Proc. Natl. Acad. Sci. USA **77**, 186-191

3. Ruvkun, G.B. and Ausubel, F.M. (1980). Proc. Natl. Acad. Sci. **77**, 191-195

4. Ruvkun, G.B. and Ausubel, F.M. (1981). Nature **289**, 85-88

5. Hennecke, H. (1981). Nature **291**, 354-355

6. Quiviger, B., Franche, C., Lutfalla, G., Rice, D., Haselkorn, R. and Elmerich, C. (1982). Biochimie **64**, 495-502

7. Merrick, M., Filser, M., Dixon, R., Elmerich, C., Sibold, L. and Houmard, J. (1980). J. Gen. Microbiol. **117**, 509-520

8. Loenen, W.A. and Brammar, W.J. (1980). Gene **10**, 249-259

9. Maniatis, T., Fritsch, E.F. and Sambrook, J. (1982). Molecular Cloning, CSH 1982

10. Hohn, B. (1979). Methods Enzymol. **68**, 299-309

11. Benton, W.D. and Davis, R.W. (1977). Science **196**, 180-182

12. Southern, E. (1975). J. Mol. Biol. **98**, 503-517

SITE-DIRECTED TRANSPOSON MUTAGENESIS OF CLONED nif-GENES
OF AZOSPIRILLUM BRASILENSE

M. Singh and W. Klingmüller
Lehrstuhl für Genetik, Universität Bayreuth,
Universitätsstraße 30, D-8580 Bayreuth, FRG

Introduction

Free-living, nitrogen fixing, bacteria (specially Azospirillum), inhabiting the rhizosphere of plants, have drawn considerable interest because of their potential application as biofertilisers. Efforts are being made to develop the genetics of Azospirillum, e.g., there have been reports of (1) chromosome mobilisation (1), (2) transfer of Inc P group plasmids as carriers of transposons (2, 3) and (3) identification of indigenous, large and small plasmids (3, 4, 5). Of particular interest is the cloning of a nif-DNA fragment from total DNA of A. brasilense ATCC 29145 (6).

For studying organisation and function of nif-genes in Azospirillum, it is necessary to obtain well defined mutants which are normal for the rest of their genome. Site-directed mutagenesis (7) is a very useful method in this respect, particularly where the desired DNA fragment has already been cloned. For systems where no transformation is possible, site-directed mutagenesis consists of the following steps, (1) transposon mutagenesis of the cloned DNA (this, in itself, is important for studying organisation and expression of the cloned genes in E. coli minicells), (2) localisation of transposon insertions, (3) back transfer of the mutagenised DNA fragment into the wild type strain, and (4) selection for defective phenotype (in our case, the nif⁻ phenotype) resulting from the genetic exchange of the wild type gene with the mutagenised one.

The nif-DNA fragment of A. brasilense has been cloned on the basis of interspecies homology with nitrogenase structural genes of K. pneumoniae (6, 8). In order to study the organisation of nif-genes in Azospirillum and to confirm that the cloned nif-DNA is functional (recently, complementation of a nif$^-$ mutant with the cloned nif-DNA has been shown, 9), we have undertaken to carry out site-directed mutagenesis of the cloned nif-genes of A. brasilense. In the present report, the results of transposon mutagenesis are described.

Bacterial strains and plasmids

Bacterial strains and the plasmids used in the present study are listed in Table 1.

Table 1

Bacteria/Plasmids	Relevant characteristics
E. coli	
HB101	Str^r, r^- m^-, $recA^-$
HB101 (pAB1)	
RU 2901 (Rts 1::Tn1725)	Nal^r, r^- m^-
Plasmids	
pAB1	Tc^r (contains a 6.7 kb nif-gene fragment of A. brasilense ATCC 29145)
Rts 1::Tn1725	Kan^r, Cm^r

Transposon mutagenesis

Transfer of Rts1::Tn1725 into HB101 (pAB1)

The plasmid Rts1::Tn1725 is a kanamycin resistant, self-transmissible, temperature sensitive replicon, and hence can be eliminated from cells by temperature treatment, thus making possible the identification of transpositions of Tn1725 (chloroamphenicol resistant) (10). The strain RU2901 (Rts1::Tn1725) was conjugated with HB101 (pAB1) at 30° C for 2 hrs and the conjugation mixture was plated on LB + kanamycin + chloroamphenicol + streptomycin. 45 transconjugants were purified and then incubated at 42° C (on plates containing chloroamphenicol and tetracycline) to eliminate Rts1. The loss of the plasmid Rts1 was verified by testing for the resistance to kanamycin in 40 isolates.

Detection of cells containing pAB1 with Tn1725 insertions

Retention of chloroamphenicol resistance following the elimination of Rts1::Tn1725, indicated the transposition of Tn1725, which could be either in the chromosome or in pAB1. In order to identify the isolates containing pAB1::Tn1725, plasmid DNA was isolated from the 40 isolates by a rapid method (11), and used to transform HB101. The transformants were selected on LB containing tetracycline and chloroamphenicol. Desired transformants were obtained in 16 cases.

Restriction analysis of Tn1725 insertions in pAB1

Because the transposon Tn1725 (8.9 kb) contains two Eco RI sites in its terminal repeats (10), digestion of pAB1::Tn1725 should produce a new band of about 8.9 kb. Plasmid DNA was isolated from the above mentioned 16 transformants and after digestion with Eco RI, it was electrophoresed on agarose gel. In each case, there was a new band of about 8.9 kb showing the insertion of Tn1725 in pAB1. This has been conclusively demonstrated by Southern hybridisation (17) using Tn1725 as the ^{32}P-labelled probe (data not shown). Eco RI digestion of pAB 1 produces two fragments, a 6.7 kb fragment containing nif-genes of A. brasilense and a 4.0 kb fragment corresponding to the vector pACYC184 (see fig. 1). Due to the insertion of the transposon Tn1725, the corresponding fragment should disappear from the Eco RI digests of pAB1::Tn1725. Eco RI digestion of plasmid DNA from 6 isolates (fig. 2 A, B and fig. 3 A, B, C, G) show that the 6.7 kb fragment has disappeared, simultaneously two new fragments have appeared showing the insertion of Tn1725 in the nif-gene fragment of pAB1. On the other hand, in other 8 isolates, the 4.0 kb fragment is missing (fig. 2 C, E, F, G and fig. 3 D, F, H, I) which locates the insertion of the transposon in the vector pACYC184; these insertions must lie outside the tetracycline gene because the isolates are still resistant to tetracycline.

In rest of the isolates (fig. 2 D and 3 E), there are only three bands visible while the 6.7 kb band is missing. The intensity of the 4.0 kb band is more than expected indicating that it is a doublet. In other words, one of the two Eco RI fragments produced due to the insertion of the transposon in the 6.7 kb nif-gene fragment, is 4.0 kb large which runs together with the vector pACYC184 and hence only three bands are visible on gels.

Constitution of plasmid pAB1

Quiviger et al. (6) cloned a 6.7 kb Eco RI fragment of
A. brasilense ATCC 29145 in the vector λ-gt 7-ara6 (on the basis
of DNA homology with K. pneumoniae nif HDKY) and subsequently
recloned the 6.7 kb fragment in the unique Eco RI site of the
vector pACYC184. The resulting recombinant plasmid was designated as pAB1. The physical map of pAB1 is shown in fig. 1.

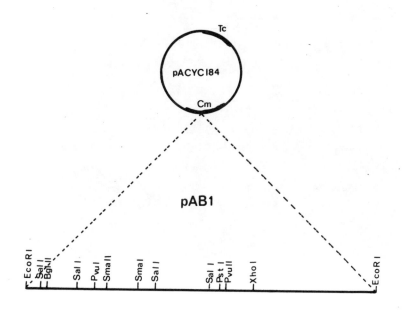

Figure 1: Physical map of plasmid pAB1; it contains a 6.7 kb
Eco RI fragment (nif-DNA) from A. brasilense, cloned
in the chloroamphenicol gene of pACYC184 (4.0 kb).
After Quiviger et al., 1982 (6).

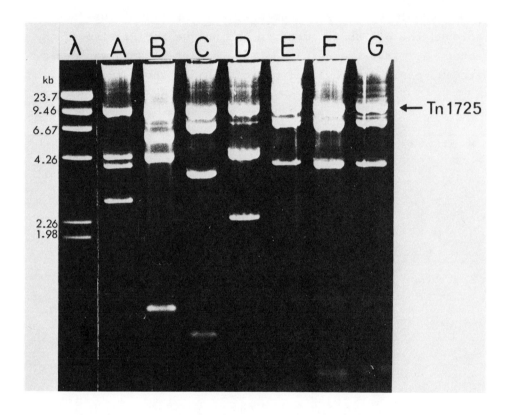

Figure 2: Agarose gel electrophoresis of Eco RI digests of plasmid DNAs from 7 chloroamphenicol resistant transformants (transpositions). λ- (left most track) refers to the Hind III fragments of λ-DNA used as length standards. Fragments larger than 10 kb in lanes E and G are partial digestion products.

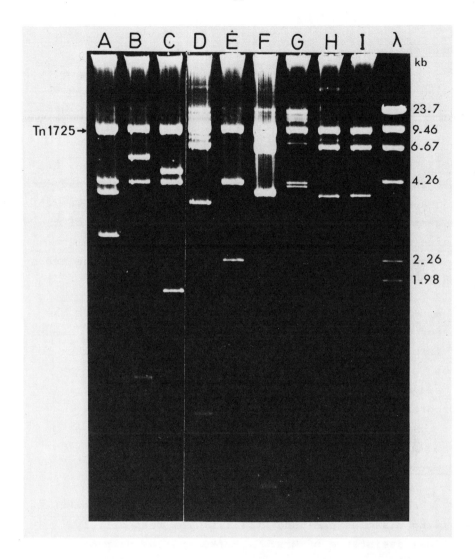

Figure 3: Agarose gel electrophoresis of Eco RI digests of plasmid DNAs from 9 chloroamphenicol resistant transformants (transpositions). λ-(right most track) refers to the Hind III fragments of λ-DNA used as length standards. Fragments larger than 10 kb are partial digestion products.

Discussion

In the present report, we have described the results of transposon mutagenesis of nif-genes of A. brasilense cloned on the basis of non-genetic criteria. When plasmid DNA from 40 presumptive transpositions was used to transform E. coli HB101, transformants were obtained in 16 cases. All the 16 isolates possessed a plasmid larger than pAB1 and each showed a new band (about 8.9 kb which is the size of Tn1725) on digestion of their plasmid DNA with Eco RI. Eight of the isolates contained the transposon insertions at different locations in the 6.7 kb nif-fragment, the rest had the transposon in the 4.0 kb fragment (vector pACAC184).

The next step in site-directed mutagenesis involves the transfer of the mutagenised genes (nif-genes::Tn1725 insertions) back into the wild type strain of A. brasilense and then to select for the resulting nif$^-$ mutants. For this, cloning of the mutagenised nif-genes fragment in a suitable vector is in progress.

Acknowledgements

I am thankful to Dr. C. Elmerich and Prof. R. Schmitt for providing plasmids pAB1 and Rts1::Tn1725, respectively. Technical assistance of Mrs. C. Refke and the help of Mr. R. Fahsold in this work is greatly acknowledged. This work was supported by the Bundesministerium für Forschung und Technologie.

References

1. Franche, C. and Elmerich, C. (1981). FEMS Microbiol. Letters 10, 199-202.

2. Singh, M. (1982). In Klingmüller, W. (ed.): Azospirillum, Genetics Physiology Ecology, EXS 42, pp 35-43, Birkhäuser, Basel.

3. Elmerich, C. and Franche, C. (1982). In Klingmüller, W. (ed.): Azospirillum, Genetics Physiology Ecology, EXS 42, pp 9-17, Birkhäuser, Basel.

4. Singh, M. and Wenzel, W. (1982). In Klingmüller, W. (ed.): Azospirillum, Genetics Physiology Ecology EXS 42, pp 44-51, Birkhäuser, Basel.

5. Wood, A.G., Menezes, E.M., Dykstra, C. and Duggan, D.E. (1982). In Klingmüller, W. (ed.): Azospirillum, Genetics Physiology Ecology, EXS 42, pp 18-34, Birkhäuser, Basel.

6. Quiviger, B., Franche, C., Lutfalla, G., Rice, D., Haselkorn, R. and Elmerich, C. (1982). Biochimie 64, 495-502.

7. Ruvkun, G.B. and Ausubel, F.M. (1981). Nature 289, 85-88.

8. Ruvkun, G.B. and Ausubel, F.M. (1980). Proc. Natl. Acad. Sci. USA 77, 191-195.

9. Jara, P., Quiviger, B., Laurent, P. and Elmerich, C. (1983). Can. J. Microbiol. (In Press).

10. Altenbuchner, J., Schmid, K. and Schmitt, R. (1983). J. Bacteriol. 153, 116-123.

11. Kado, C.I. and Liu, S.T. (1981). J. Bacteriol. 145, 1365-1373.

12. Southern, M. (1975). J. Mol. Biol. 98, 503-517.

UPTAKE AND FATE OF PLASMID-DNA IN AZOSPIRILLUM LIPOFERUM A23

Gerhard Schwabe and Ulrike Weber

Lehrstuhl für Genetik, Universität Bayreuth,
Universitätsstraße 30, D-8580 Bayreuth, FRG

Introduction

Genetic transformation has been and is still an important way of looking at the chromosome of bacteria, mapping their genes, and studying the function of particular genes. For experiments concerning plasmid-DNA and cloned genes, known as genetic engineering, transformation has gained particular importance. The bacteria of the genus Azospirillum have recently attracted much interest because of their ability to fix nitrogen and to synthesize plant hormones, for example auxin. In spite of many efforts Azospirillum can not uptil now be transformed by plasmids (1).

Generally the process of transformation by plasmids as well by chromosomal DNA is well understood for many bacteria (2, 3). It can be devided into the steps of
a) binding of the plasmid to the recipient cell,
b) uptake of the plasmid into the cell,
c) establishing of the plasmid,
d) replication and expression of the plasmid.
Step (d) of replication and expression of foreign plasmids can be carried out by Azospirillum, as successful conjugation experiments have shown (4, 5).

Here we report on the results of experiments to find out whether Azospirillum can take up a foreign plasmid and to follow the fate of this DNA once it is inside the cell.

Material and Methods

Bacteria and plasmids: As recipients for the transformation experiments <u>Azospirillum lipoferum</u> A23 (isolated from maize by G. Jagnow, Braunschweig) and <u>Escherichia coli</u> RRI were used. The bacteria were made competent by $CaCl_2$-treatment (6). As a donor the plasmid pUW 610 (1, formerly named pBT 610) was used. This plasmid is a hybrid between pACYC 177 and a Hind II-fragment of pBR 325; it confers resistance to the antibiotics chloramphenicol and kanamycin. The plasmid was isolated from E. coli lysates by two successive CsCl-gradient ultracentrifugations and was labelled with ^{3}H-dCTP by nick translation (7).

Incubations: 120 µl competent cells were incubated with 10 µl ^{3}H-pUW 610 for 20 min. on ice, and then for 5 min. at 30° C (Azospirillum) or 37° C (E. coli). 0.5 ml LB-medium was added, and the cells were incubated for additional 30 min. at 30°/37° C.

For DNase-treatment the cells were centrifuged and resuspended in 1 ml LB, 10 mM $MgCl_2$ with 5 µg DNase and incubated for 15 min. at 37° C. The bacteria were collected and lysed by the method of Kado and Liu (8), without phenol extraction. The plasmid-DNA was precipitated with ice-cold trichloracetic acid (TCA) on filters. The radioactivity in the precipitant was determined in a liquid scintillation counter. The radioactivity was used to quantitate the amount of plasmid-DNA.

For testing the stability, the bacteria were collected after 30 min. incubation and resuspended in 1 ml fresh LB-medium and incubated at 30°/37° C. At various times 100 µl samples were taken and treated with DNase, lysed, and the DNA was precipitated as described above.

Mutagenization of A. lipoferum A23 and screening for mutants with a reduced nuclease activity: The sediment of 9.5 ml overnight culture of A. lipoferum A23 was resuspended in 10 ml tris-malate buffer with 100 µg/ml nitrosoguanidin. The bacteria were incubated for 30 min. at room temperature, washed 3 times with LB-medium and plated on LB-agar.

The screening for mutants with a reduced nuclease activity was done according to Dürwald and Hoffmann-Berling (9): After mutagenesis the colonies were replica-plated on complete medium in glass petri dishes. The colonies on the glass plates were treated with toluol (15 ml toluol/plate, 4 hrs at 30° C), dryed and stained with 3,5 % (v/v) giemsa in 0.1 M Na_2HPO_4, pH 7.4, 20 % (v/v) ethanol for 20 min. at room temperature. The colonies were destained with the Na_2HPO_4/ethanol buffer.

Nuclease activity assay: Cell extracts were prepared from 2 ml overnight cultures. The cells were centrifuged and resuspended in 0.5 ml 10 mM Tris-HCl, pH 7.4, 10 mM EDTA, 10 mM 2-mercaptoethanol, 0.5 mg/ml lysozyme. The solution was frozen, thawn, sonified for 15 sec. and centrifuged.

The assay mixture consisted of 10 µl pUW 610 and 25 µl cell extract containing 5 µg protein (10), in 10 mM $MgCl_2$. As a control the plasmid was incubated with buffer. It was incubated at 30° C for 20 min. The incubation was stopped by adding 4 M urea and heated at 70° C for 15 min. The assay mixture was loaded on a agarose gel (1 % agarose in tris/phosphate/EDTA with 1 µg/ml ethidiumbromide) and electrophorised.

After washing, 3.8 % of the DNA was found associated with the cells and 96.6 % in the supernatant of the washing. DNase-treatment reduces the amount of DNA in the bacterial sediment to 1.8 %. Thus approximately 4 % of the plasmid-DNA in the reaction mixture is bound to the cells of Azospirillum but only 2 % are taken up into the cells. For comparison, E. coli RRI takes up 10-13 % of the DNA, i.e., at least 5 times more than Azospirillum.

In the next set of experiments the stability of the foreign plasmid in Azospirillum and as a comparison in E. coli was tested. Radioactively labelled pUW 610 and competent cells were incubated as described above. After 30 min. incubation the cells were centrifuged and resuspended in fresh LB-medium and incubated for additional 3 hrs.

Table 2: Stability of plasmid DNA which was taken up into A. lipoferum A23 and E. coli RRI

^{3}H-DNA in the reaction mixture	A23 195 356 cpm = 100 %	RRI 197 806 cpm = 100 %
	TCA-precipitated radioactivity from cell lysates after incubation	
Incubation time		
0 hrs.	2 126 cpm = 1.1 %	12 676 cpm = 6.4 %
0.5 hrs.	1 128 cpm = 0.6 %	10 773 cpm = 5.4 %
1 hr.	1 117 cpm = 0.6 %	7 392 cpm = 3.7 %
2 hrs.	1 584 cpm = 0.8 %	7 820 cpm = 3.9 %
3 hrs.	1 174 cpm = 0.6 %	8 265 cpm = 4.2 %

At times indicated in table 2 samples were taken, treated with DNase, lysed, and prepared for TCA-precipitation. Azospirillum has taken up only 1.1 % of the DNA present in the reaction mixture. After 0.5 hrs. of incubation, the amount of plasmid decreased to about one half of the original value and then remained constant. In E. coli there is only a gradual reduction to approximately 2/3 of the original value at the end of the incubation of 3 hrs.

When considerung these results, it is necessary to note, that E. coli has a higher growth rate than Azospirillum. After the $CaCl_2$-treatment, Azospirillum needs a much longer time for the first doubling of the cell titre as compaired to E. coli.

On the basis of the upper results we conclude that, due to the limited uptake of plasmid-DNA into the cells of Azospirillum and due to the rather effective degradation of the plasmid-DNA inside the cells by nucleases present in Azospirillum, the amount of plasmid-DNA inside the cells decreases below a critical value, so that the plasmid is not able to establish itself.

One way to circumvent this problem would be, to make mutants with a reduced nuclease activity. For this A. lipoferum A23 was mutagenised with nitrosoguanidine and screened for the appropiate mutants (9): The replica-plated colonies were treated with toluol to initiate uncontrolled breakdown of the DNA and then were stained with giemsa. In cells with normal nuclease activities the cellular DNA is degradaded, but in mutants with a reduced nuclease activity the cellular DNA is not degradaded and consequently these colonies are blue after giemsa staining. Nine such colonies were picked from which the nuclease activity was determined. Cell extracts were made and incubated with pUW 610 for 20 min. The reaction mixture was analysed by agarose gel-electrophoresis. The intensity of the ethidiumbromide-stained plasmid bands was used to quantitate the plasmid-DNA. Fig. 1 shows that mutant no. 8 has a significantly reduced nuclease

Results and Discussion

During these investigations the plasmid pUW 610 was used as a model system. It was radioactively labelled by nick translation. The recipient cells were from A. lipoferum A23 and as a reference E. coli RRI, which is commonly used in transformation experiments. The recipient cells were treated with $CaCl_2$ and then incubated with ^{3}H-pUW 610 under various conditions.

First it was investigated, whether Azospirillum can take up plasmid-DNA or whether the DNA is only unspecifically bound to the cell. After incubation with ^{3}H-pUW 610 the cells were washed 3 times. In a parallel experiment the cells were treated with DNase and washed (table 1).

Table 1: Binding or uptake of plasmid-DNA into A. lipoferum A23

^{3}H-DNA in the reaction mixture	33 9000 cpm = 100 %	
	TCA-precipitated radioactivity in the	
	bacterial sediment	bacterial supernatant
cells washed 3x	1 320 cpm = 3.8 %	32 792 cpm = 96.9 %
cells treated with DNase and washed 3 x	690 cpm = 1.8 %	2 070 cpm = 6.1 %

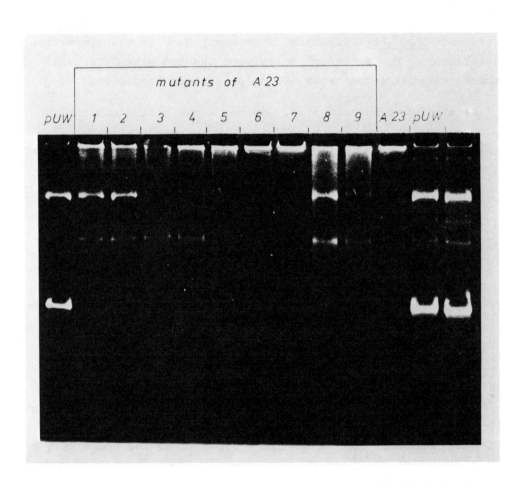

Figure 1: Nuclease activity in extracts of A. lipoferum A23 and mutants no. 1-9.

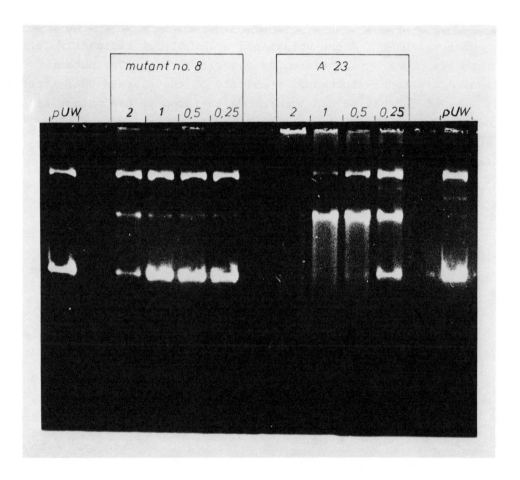

Figure 2: Estimation of the nuclease activity in extracts
of A. lipoferum A23 and mutant no. 8.
In the reaction mixture different amounts of cell
extract with 2, 1, 0.5, and 0.25 µg protein were
incubated with pUW 610.

activity as compared to the parental strain A23. A further test revealed (fig. 2), that approximately 4 times more protein of mutant no. 8 is necessary to obtain the same nuclease activity as in A23.

For mutant no. 8 the uptake and stability of radioactively labelled pUW 610 were tested as described above. The mutant showed a slightly increased stability of the plasmid-DNA: After 4 hrs. of incubation 0.5 % more DNA is found in the mutant as compared to that in the parental strain. This increase of stability, however, is not sufficient for a successful transformation, as further experiments have shown.

References

1. Weber, U. (1982). Diplomarbeit, Universität Bayreuth.

2. Smith, H.O., Danner, D.B. and Deich, R.A. (1981).
   Ann. Rev. Biochem. 50, pp 41-68.

3. Humphreys, G.O., Weston, A., Brown, N.G.M. and Saunders
   J.R. (1978). In Glover, S.W. and Butler, L.O. (eds.):
   Transformation 1978, pp 287-312, Cotswold, Oxford.

4. Polsinelli, M., Baldani, E., Bazzicalupo, M. and Gallori, E.
   (1980). Molec. gen. Genet. 178, pp 709-711.

5. Singh, M. (1982). In Klingmüller, W. (ed.): Azospirillum,
   Genetics Physiology Ecology, EXS 42, pp 35-43.

6. Cohen, S.N. and Chang, A.C.Y. (1972). Proc. Natl. Acad.
   Sci. USA 69, pp 2110-2114.

7. Rigby, P.W.J., Dieckmann, M., Rhodes, C. and Berg, P. (1977).
   J. mol. Biol. 113, pp 237-251

8. Kado, C.I. and Liu, S.-T. (1981). J. Bact. 145, pp 1365-1373.

9. Dürwald, H. and Hoffmann-Berling, H. (1968). J. molec. Biol.
   34, pp 331-346.

10. Lowry, O.H., Rosebrough, N.J., Farr, A.L. and Randahl, R.J.
    (1951). J. biol. chem. 193, pp 265-275

NIF MUTANTS OF AZOSPIRILLUM BRASILENSE:
EVIDENCE FOR A NIF A TYPE REGULATION

F.O. PEDROSA* and M.G. YATES

ARC Unit of Nitrogen Fixation, University of Sussex
Brighton, BN1 9RQ, UK., and *Department of Biochemistry
Universidade Federal do Parana, C. Postal 939,
80000, Curitiba, PR, Brazil

Introduction

Nitrogenase genes (nif) expression in Klebsiella pneumoniae is regulated by a cascade system involving nifA, nifL, ntrA (glnF), ntrC (glnG) and ntrB (glnL): the ntrA and ntrC gene products together activate the nifLA operon and the nifA product activates the other nif promotors in an ntrA background. The ntrB and ntrC gene products together repress the ntr system and the nifL gene product represses nif expression. This latter repression is triggered by NH_4^+ or O_2 (see 1). The ntr genes control expression of several other N-utilising systems including proline, histidine, arginine and NO_3^-. The nifA gene product can substitute for the ntrC gene product in this system (1,2).

We have isolated and characterised Nif⁻ mutants of Azospirillum brasilense and present evidence for the involvement of nifA and the ntr system in the regulation of nif expression in this organism.

Materials and Methods

Organisms and plasmids (Table 1). A. brasilense Sp7 Nal^{r}15, Sm^{r}200 (FP2) is a naturally occurring mutant of Sp7 (ATCC 29145). All Nif⁻ mutants were obtained from FP2 following N-nitrosoguanidine mutagenesis.

Plasmid pCK3, a cointegrate of pRK290 and an EcoRl DNA fragment from pMC71a (3) containing K. pneumoniae nifA expressed constitutively from a tetracycline promotor was provided by Dr. C. Kennedy.

Table 1: Strains and plasmids used

Strains	genotype/phenotype	source or reference
A. brasilense Sp7		
(ATCC 29145)	nif	Tarrand et al. (1978)
FP2	nif Nal^r/Sm^r	This work
FP3,4,5,6,7 8,9,10	nif^- Nal^r/Sm^r	"
Escherichia coli		
HB101	pro^-	C. Kennedy
5K	$thr^- leu^- thi^-$	"
1230	$pro^- met^-$	"
Plasmids		
pCK3	K. pneumoniae $nifA^c$, Tc^r, inc P-1	"
R68.45	Km^r, inc P-1	"
pRK2013	Tra, Km^r	"

Azospirillum strains were grown in NfbHP medium, a modification of Nfb (4) containing in g/L: K_2HPO_4: 5, KH_2PO_4: 4, $FeSO_4 \cdot 7H_2O$: 0.020, nitrilotriacetic acid: 0.056, sodium lactate: 5, with either N_2, NH_4Cl (20 mM) or sodium glutamate (5 mM) as N source. Escherichia coli 5K (pCK3) and 1230 (R68.45) were grown on LB Tc^5 (5) and HB101 (RK2013) on LB Km^{15}. Plasmid pCK3 required mobilising by plasmid RK2013, so the E. coli donors and A. brasilense were mated as trios on LB-NfbHP-NH_4Cl (1:1) plates at 30° for 24 h. The transconjugants were mated with E. coli 1230 (R68.45) under similar conditions to eliminate plasmid pCK3. A. brasilense (pCK3 or R68.45) transconjugants were selected in solid NfbHP-NH_4Cl Nal^{15} Tc^{10} or Nal^{15} Km^{60} respectively.

Mutagenesis and isolation

FP2 was mutagenised with NTG (20 µg/ml; 50% survivors, outgrown for 10 generations in NfbHP-NH_4Cl and selected for poor growth on solid NfbHP under O_2:CO_2:N_2 (1:1:98). Small colonies were tested for Nif in either semi-solid NfbHP (1 ml) with no N, 1 mM NH_4Cl or 5mM glutamate or liquid NfbHP-glutamate under air (100 ml in 250 ml conical flasks at 100 rpm and 30°).

Biochemical

Nitrogenase fractions, the MoFe component (Ab1), the Fe component (Ab2) and the Fe component activating factor (Af) were isolated and partly purified by DEAE cellulose chromatography of a crude supernatant fraction from disrupted cells (6). Complementation studies with Nif⁻ mutant crude extracts, obtained by disruption in the French pressure cell, were by dithionite-dependent assay (6) with 13% C_2H_2 for 15 min at 30°.

In vivo nitrogenase activity was determined by C_2H_2 (10%) reduction. C_2H_4 was measured by gas chromatography with a Poropak N column. Glutamine synthetase (GS) and glutamate synthase were determined as described (7,8).

Plasmid patterns were determined by the 'boiling' procedure (9) and protein by the Folin method with bovine serum albumin as the standard.

Results

Nif mutants

237 (25%) of the small colonies selected from N-free plates showed less than 10% of wild type nitrogenase activity in N-free semi-solid medium at 37° but only 8 of these isolates had <10% wild type nitrogenase activity in liquid NfbHP-glutamate under air at 30°. Addition of glutamate made O_2 precautions unnecessary in liquid medium and these conditions proved more effective for nif expression than the standard N-free semi-solid medium. Table 2 describes a typical growth experiment with glutamate under air. Four of the eight mutants isolated had no nitrogenase activity, 3 had <5% and 1 <10% of wild type

activity (Table 3).

Table 2: Growth and nitrogenase activity of A. brasilense FP2 in glutamate-containing medium

Culture turbidity at A540 nm EEL units	Protein Concentration mg/ml Culture	Nitrogenase Activity	
		nmol C_2H_2/ min.ml culture	nmol C_2H_2/ min.mg protein
8	0.06	zero	zero
12	0.06	zero	zero
15	0.08	zero	zero
26	0.22	0.2	0.9
31	0.26	0.9	3.4
34	0.29	4.9	16.7
40	0.39	5.5	14.1
42	0.39	5.4	13.8
51	0.45	5.9	13.1

Bacteria were grown in liquid medium containing glutamate and assayed "in situ" as described under Materials and Methods.

Table 3: Isolation of Nif⁻ mutants of A. brasilense after NTG treatment

	Number	Frequency
Mutagenized cultures	30	-
Colonies examined in N-free plates	21212	1
Small colonies tested for Nif⁻	787	3.7×10^{-2}
Isolates showing <10% of wild type nitrogenase activity in semi-solid medium at 37°C	224	1.1×10^{-2}
Isolates showing <10% of wild type nitrogenase activity in glutamate medium	8	3.8×10^{-4}

% Activity	Mutant designation
0	FP3,8,9,10
0-5	FP5,6,7
5-10	FP4

Procedures were described under Materials and Methods.

Growth on N sources

All the mutants except FP5 grew as rapidly as the wild type on NfbHP containing 20 mM NH_4Cl (doubling time ~115 min). FP3, 4 and 5 grew poorly with 1 mM NH_4Cl or histidine, FP4 grew poorly and FP5,8 and 9 failed to grow on NO_3^-. FP5 grew poorly but FP6, 7 and 10 grew normally (as FP2) on all N sources tested (20 mM and 1 mM NH_4Cl, histidine, arginine, proline and NO_3^-) (Table 4).

Table 4: Growth of Nif⁻ mutants of *A. brasilense* FP2 on different N-sources

Strain	NH_4Cl 1mM	NH_4Cl 20mM	Arg 1 mg/ml	Pro 1 mg/ml	NO_3^- 10mM	His (1) 1 mg/ml	Glu 5mM
FP2,6,7,10	+	+	+	+	+	+	+
FP8,9	+	+	+	+	− (2)	+	+
FP3	poor	+	+	+	+	poor	+
FP4	poor	+	+	+	poor	poor	+
FP5	poor	poor	poor	poor	−	poor	poor

Growth in solid NfbHP plus N-sources at 30°C for 48 h.
(1) 72 h
(2) pCK3 transconjugants of FP8 and 9 grew normally on NO_3^-.

Biochemical Studies

Nitrogenase

Crude extracts of FP3, 8, 9 and 10 showed no nitrogenase activity *in vitro* nor were they complemented by either Ab1 or Ab2. FP6 contained low levels of Ab1 but over-produced Ab2. FP4, 5 and 7 had low levels (<10% of FP2) of both Ab1 and Ab2. All the mutants contained normal amounts of Ab2-activating factor (Table 6).

Typical experimental results are shown in Table 5: Ab1 and Ab2-activating factor showed no nitrogenase activity; Ab2 was contaminated with Ab1 and activity was enhanced 3-fold by adding the activating factor. FP3 showed activity only when complemented by Ab1 plus Ab2 indicating the presence of Ab2-activating factor. FP6 was complemented by Ab1 and, to a small extent, by Ab2.

Table 5: Biochemical complementation

	Nitrogenase activity (nmol C_2H_4/15 min)			
Addition \ Strain	None	FP3	FP6	FP2 (wild-type)
None	0.0	0.0	4.8	30.0
Ab1	0.0	0.0	275.6	37.5
Ab2	5.1	20.4	38.6	110.0
Ab1 + Ab2	18.9	159.7	420.9	186
Af	0.0	0.0	7.4	47.6
Ab1 + Af	0.0	0.0	256.9	56.6
Ab2 + Af	16.5	24.8	49.4	161.4
Ab1 + Ab2 + Af	132.2	154.9	380.9	238.8

Reaction mixture modified from Ludden et al. (6).
Nitrogenase activity was determined in crude extracts as described under Materials and Methods.   Protein concentrations (mg) were: Ab1 1.15; Ab2 0.25; Af 2.3; crude extracts 0.5 to 2.

Glutamine synthetase and glutamate synthase

FP8 and 9 had <30% of wild type GS activity (transferase assay) and FP8 had no glutamate synthase. All other mutants had similar activities to the wild type (Table 6).

Table 6: Biochemical characterizations of Nif⁻ mutants of
A. brasilense FP2

Nitrogenase	FP2	3	4	5	6	7	8	9	10
Ab1	+	-	tr	tr	tr	tr	-	-	-
Ab2	+	-	tr	tr	++	tr	-	-	-
Ab2 activating factor	+	+	+	+	+	+	+	+	+
Glutamine Synthetase (Mn^{2+})	+	+	+	+	+	+	<30%	<30%[1]	+
Glutamate Synthase	+	+	+	+	+	+	-	+	+

Determinations described under Materials and Methods.

Wild-type GS and GOGAT activities were 6.2 and 0.039 μmol product/min. mg protein

(1) % of wild type activity

Genetical complementation

Nitrogenase activity. Plasmid pCK3 was mobilised by plasmid pRK2013 into strains FP2 to 10 at 10^{-3} per recipient; transfer was confirmed by plasmid electrophoresis on agarose gels. FP8, 9 and 10 transconjugants showed nitrogenase activity at 30° but not at 37°; consistent with the known thermolability of the K. pneumoniae nifA gene product (3)(Table 7). Nitrogenase activities of the pCK3 transconjugants of FP8, 9 and 10 were abolished by introducing plasmid R68.45 of the same P.1 incompatibility group (Table 1, 7).

Plasmid pCK3 transconjugants of FP2, 8, 9 and 10 expressed nitrogenase activity with 20 mM NH_4Cl (Table 8). Some activity was observed "in situ" (undisturbed growing cultures) though less than "in vitro" (crude extracts of NH_4-grown transconjugants) indicating some inhibition of activity by NH_4 in vivo. Plasmid R68.45 eliminated constitutive nif expression (Table 8).

Growth on NO_3^-. FP5, 8 and 9, pCK3 transconjugants grew as well as the wild type on NfbHP plus nitrate as the sole N source. N-free controls failed to grow indicating that NO_3^- was assimilated by the transconjugants.

Table 7: Genetical complementation of **A. brasilense** FP2 Nif⁻ mutants by pCK3(nifA^c) and elimination by R68.45

Strain	Nitrogenase activity (nmol C_2H_4/min.mg protein)	
	30°C	37°C
FP8	0	0
FP8 (pCK3)	6.5	0
FP9	0	0
FP9 (pCK3)	6.5	0
FP10	0	0
FP10 (pCK3)	8.4	0
FP10 (R68.45)*	0	0
FP2	25.7	27.9
FP2 (pCK3)	20.9	10.8
FP2 (R68.45)*	26.6	11.9

Bacteria were grown in N-free semi-solid NfbHP medium at 30° and 37°C. Nitrogenase activity was assayed at the growth temperature. *After elimination of plasmid pCK3 by plasmid R68.45

Table 8: Constitutive expression of nitrogenase activity in **A. brasilense** strains carrying plasmid pCK3(nifA^c)

Strain	Nitrogenase activity (nmol C_2H_4/min.mg protein)	
	in situ (1)	in vitro (2)
FP2	0.0	0.0
FP2 (pCK3)	0.3	0.9
FP2 (R68.45)	0.0	0.0
FP8	0.0	0.0
FP8 (pCK3)	1.5	2.4
FP9	0.0	0.0
FP9 (pCK3)	1.9	4.2
FP10	ND	0.0
FP10 (pCK3)	ND	0.2
FP10 (R68.45)	ND	0.0
FP2 (glutamate grown)	8.4	8.3

Cultures were grown in NfbHP medium containing 20 mM NH_4Cl and Tc^5. Assays were as described under Materials and Methods.

(1) in situ - in growing undisturbed culture
(2) in vitro - in crude extracts

Switch-off of nitrogenase activity by NH_4^+

Adding 1 mM NH_4Cl completely switched off nitrogenase activity in vivo in glutamate-grown FP2 or in the pCK3 transconjugants of FP2 and FP10 but only caused 25 and 29% inhibition in FP8 or FP9 transconjugants (Table 9). This failure to switch off may be related to the low GS levels of these mutants.

Table 9: "Switch off" of nitrogenase activity by NH_4Cl

Strain	Nitrogenase activity (nmol C_2H_4/min.mg protein)		% Inhibition by NH_4^+
	$-NH_4^+$	$+NH_4^+$	
FP2	9.8	0.0	100
FP2(pCK3)	12.1	0.0	100
FP8	0.0	0	-
FP8(pCK3)	11.1	8.4	24
FP9	0.0	0	-
FP9(pCK3)	9.9	7.1	29
FP10	0.0	0.0	-
FP10(pCK3)	8.9	0.0	100

Bacteria were grown in NfbHP-glutamate medium and assayed in vivo as described under Materials and Methods. C_2H_2 (10%) was added at zero time and C_2H_4 determined at 60 min when 1 mM NH_4Cl (in 0.1 ml) was added. C_2H_4 was assayed again at 30 and 60 min after NH_4^+ additions.

Discussion

Eight Nif^- mutants of A. brasilense FP2 were obtained in this study. Three of these, FP8, 9 and 10, had no detectable nitrogenase activity in vivo, showed neither Ab1 nor Ab2 activity in vitro and were complemented in vivo by K. pneumoniae nif A. That this complementation was due to nifA was emphasised by the temperature sensitivity and constitutive nature of nitrogenase activity (3) in the transconjugants and its elimination by plasmid R68.45 of the same incompatibility group as plasmid pCK3. FP10 grew well on all fixed N sources tested and had wild type levels of glutamine synthetase and glutamate synthase which suggests that it is a nif A⁻ type mutant. FP8 and 9, on the other hand, failed to grow on NO_3^- whereas the pCK3 transconjugants grew normally. Since the nifA gene product can substitute for the ntrC gene product in K. pneumoniae to promote growth on various N sources, including N_2, it is possible that FP8 and 9 are ntr C⁻ type mutants. The low level of GS activity in these strains, a feature of ntr C⁻ mutants (see 10), supports this conclusion. However, FP8 and 9 grew well on N sources other than N_2 and NO_3^- which is not consistent with a ntr C⁻ mutation. We cannot, as yet, offer an explanation for this apparent discrepancy. FP8 also lacked glutamate synthase and is phenotypically similar to the glutamate synthase negative mutants of A. brasilense described by Bani et al. (11) which also failed to grow on nitrate or fix N_2 but grew on other N sources.

FP3 also showed zero nitrogenase activity in vivo and lacked both Ab1 and Ab2 activity in vitro, but it was not complemented by K. pneumoniae nifA. It grew poorly on 1 mM NH_4Cl and histidine but well on other N sources. FP4, 5 and 7 had low levels of both Ab1 and Ab2, FP4 and 7 grew poorly on some N sources whereas FP5 grew poorly on all and not at all on NO_3^-. However the pCK3 transconjugant of FP5 showed some NO_3^--dependent growth. Further characterisation of these mutants is necessary.

FP6 had low levels of Ab1 but high Ab2 indicating that it has a structural mutation in either the sub units or FeMo cofactor of Ab1 similar to the one nif⁻ mutant of A. brasilense characterised to date (12).

We conclude that nif expression in A. brasilense is controlled by a nif A type and possibly ntr type regulatory system analogous to that in K. pneumoniae. K. pneumoniae nifA has also been shown to complement nif⁻ strains of Azotobacter vinelandii, A. chroococcum (13) and Rhizobium meliloti (14). Since A. brasilense fixes N_2 at 37° its nifA type gene product is less temperature sensitive than that of K. penumoniae.

A. brasilense nitrogenase is reversibly 'switched off' by NH_4^+ and requires an Fe protein activating factor (15) which suggests that its nitrogenase activity is regulated through an adenylylation-deadenylylation mechanism similar to that in photosynthetic bacteria (16). The pCK3 transconjugants of FP2 and FP10 'switched off' nitrogenase with 1 mM NH_4Cl whereas those of FP8 and 9 showed only a partial inhibition. This failure to 'switch off' completely may be related to low GS activity: FP8 and FP9 had only 30% of the wild type activity. FP2 and FP10 pCK3 transconjugants also expressed nif and showed low levels of nitrogenase activity in vivo when grown in the presence of NH_4^+. This again may be related to low GS activity due to repression by NH_4^+. Elimination of A. brasilense GS activity either by mutation (12) or by inhibition with methionine sulfoximine (17) also allowed constitutive nitrogenase activity in vivo in the presence of NH_4^+. All this evidence indicates a role for GS in the regulation of nitrogenase activity in A. brasilense.

The discovery that A. brasilense will grow and fix N_2 under air when supplemented with glutamate should contribute decisively to the search for further Nif⁻ mutants necessary to understand the regulation of nitrogenase synthesis and activity in this organism.

Acknowledgements

We wish to thank Dr C. Kennedy for plasmid pCK3 and Drs F. Cannon, M. Merrick, C. Kennedy, R. Dixon and R. Robson and Professor J.R. Postgate for useful discussion. F.O.P. thanks CNPq and the British Council for support.

References

1. Merrick, M. 1983. EMBO J. 2, 39-44

2. Ow, D. and Ausubel, F.M. 1983. Nature (London) 301, 307-313

3. Buchananan-Wollaston, V., Cannon, M.C., Beynon, J.L. and Cannon, F.C. 1981. Nature (London) 294, 776-778

4. Pedrosa, F.O., Dobereiner, J. and Yates, M.G. 1980. J. Gen. Microbiol. 119, 547-551

5. Miller, J.H. 1972. Experiments in Molecular Genetics, Cold Spring Harbour, New York, p 433

6. Ludden, P.W., Okon, Y. and Burris, R. 1978. Biochem. J. 173, 1001-1003

7. Goldberg, R.B. and Hanau, R. 1979. J. Bacteriol. 137, 1282-1289

8. Meers, J.L., Tempest, D.W. and Brown, L.M. 1970. J. Gen. Microbiol. 64, 187-194

9. Holmes, D. and Quigley, M. 1981. Anal. Biochem. 114, 193-197

10. Espin, G., Alvarez-Morales, A., Merrick, M. 1981. Mol. Gen. Genet. 184, 213-217

11. Bani, D., Barberio, C., Bazzicalupo, M., Favilli, F., Gallori E., Polsinelli, M. 1980. J. Gen. Microbiol. 119, 239-244

12. Gauthier, D. and Elmerich, C. 1977. FEMS Microbiol. Lett. 2, 101-104

13. Kennedy, C. and Robson, R. 1983. Nature (London) 301, 626-628

14. Sundaresan, V., Jones, J.D.G., Ow, D.W. and Ausubel, F.M. 1983. Nature (London) 301, 728-732

15. Okon, Y., Houchins, J.P., Albrecht, S.L. and Burris, R.H. 1977. J. Gen. Microbiol. 98, 87-93

16. Ludden, P. and Burris, R.H. 1979. Proc. Natl. Acad. Sci. U.S.A. 76, 728-732

17. Okon, Y., Albrecht, S.L. and Burris, R.H. 1976. J. Bacteriol. 128, 592-597

MUTANTS OF AZOSPIRILLUM AFFECTED IN NITROGEN FIXATION AND AUXIN PRODUCTION

A. Hartmann, A. Fußeder[+] and W. Klingmüller
Lehrstuhl für Genetik und Lehrstuhl für Pflanzenphysiologie[+]
Universität Bayreuth, Universitätsstr. 30, D-8580 Bayreuth, FRG

Introduction

Nitrogen fixation of Azospirillum is restricted to low concentrations of oxygen and nitrogen in the environment. In order to improve nitrogen fixation, mutants which can fix nitrogen at higher oxygen and nitrogen levels are desirable. On the other hand, the effects of Azospirillum on plant root growth are thought to be due to their hormone production. In this context, auxin overproducing mutants could particularly influence root growth.

Material and Methods

Table 1 summarizes the bacterial strains used in the present study.

Table 1: Azospirillum wild type and mutant strains

strains	source and main characteristics
A. brasilense Sp 7	ATCC no.29145, wild type strain
mutants C 1, C 2, C 3	this study, carotenoid overproducing mutants of Sp 7
A. brasilense Sp Cd	ATCC no.29710, wild type strain
mutant W 6	this study, carotenoidless mutant of Sp Cd
mutant MSX 25	this study, methionine sulfoximine (MSX) resistant mutant of Sp Cd
mutant FT 326	(1), auxin overproducing mutant of Sp Cd
A. lipoferum Sp RG 20a	ATCC no.29708, wild type strain
mutant MSX 12	this study, MSX resistant mutant of Sp RG 20a

Mutants were obtained after treatment with N-methyl-N'-nitro-
-N-nitrosoguanidine (100 µg/ml for 30 minutes at 30° C) and
growth overnight in trypton broth. The minimal media used were
prepared according to Albrecht and Okon (2). The carotenoid
contents were measured after extraction from the cells
colorimetrically according to Nur et al. (3). Ammonium concentrations were determined using the method of Fawcett and
Scott (4). The synthetic reaction of glutamine synthetase (GS)
was measured according to Kleinschmidt and Kleiner (5). Plant
growth experiments were performed in glass bottles with sterilized quartz sand (nitrogen content < 0.0005%) supplemented with
Hoagland solution to maximal water capacity. One surface sterilized maize seedling was planted per pot, inoculated with
$1 \cdot 10^8$ bacteria and incubated in a greenhouse at 20-30° C
(16 hours light, 8 hours dark).

Results and Discussion

1. Carotenoid mutants

It was suggested by Y. Okon's group, that carotenoids could
play a role in oxygen protection of nitrogen fixation in
Azospirillum (3). They described an increased oxygen sensitivity
of nitrogen fixation in carotenoidless strains as compared to
the carotenoid containing, red coloured A. brasilense Sp Cd. We
checked the carotenoid hypothesis with carotenoid negative
mutants of the strain Sp Cd and with carotenoid overproducing
mutants of the colourless A. brasilense Sp 7. After mutagenesis,
colour mutants could be found with a frequency of 10^{-2} to 10^{-3}.
Several deep red coloured mutant colonies of Sp 7 were isolated
and one mutant (C 1) was characterized in detail. After extraction of the pigments from the cells, the extracts were
chromatographed on silica gel plates, as described by Nur et al.
(3). The mutant chromatogram showed yellow and red coloured spots
at R_f-values described for the carotenoids of A. brasilense Sp Cd
(3). Fifty times concentrated extracts of the Sp 7 wild type

revealed a red coloured spot at the position of the main carotenoid of the mutant C 1. When eluted from the chromatogram, the yellow and red coloured pigments gave a carotenoid like spectrum with maximum wavelengths at 450 and 500 nm. From this it can be concluded, that the red coloured mutants of Sp 7 harboured a mutation, causing a deregulated, highly increased synthesis of carotenoids. On the other hand, white mutants of A. brasilense Sp Cd were isolated, which harbour 1 % of the carotenoid content of the wild type.

In order to demonstrate oxygen tolerance, acetylene reduction of wild type and mutant strains was tested in small volumes of liquid, nitrogen free minimal medium with air in the gas phase. In contrast to the wild type, the carotenoid overproducing mutants C 1, C 2 and C 3 showed acetylene reduction (fig. 1)

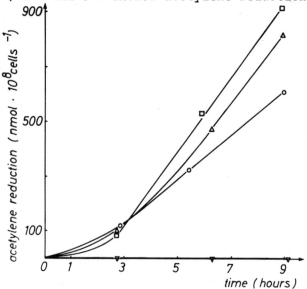

Figure 1: Nitrogen fixation (acetylene reduction) of A. brasilense Sp 7 (▽) and the carotenoid overproducing mutant C 1 (□), C 2 (○) and C 3 (△). Liquid nitrogen free minimal medium (2 ml) was inoculated with $1 \cdot 10^6$ bacteria per ml and incubated with air in the gas phase at stagnant conditions overnight at 30° C before acetylene reduction was measured.

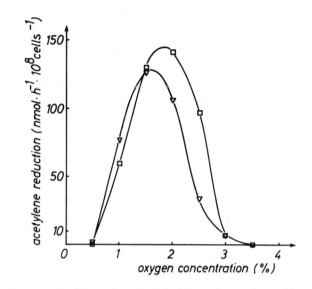

Figure 2: Nitrogen fixation activity of A. brasilense Sp 7 (▽) and the mutant C 1 (□) at different oxygen concentrations in the gas phase. Liquid nitrogen free medium was inoculated with $1 \cdot 10^7$ bacteria per ml, incubated at 37° C with shaking and acetylene reduction activity was determined after 3 hours.

and increased growth, which was determined by measuring protein and viable counts at the end of the experiment. In addition, we determined the acetylene reduction of wild type and mutant strains at different oxygen concentrations in the gas phase (fig. 2). The carotenoid overproducing mutant C 1 revealed, that the range of oxygen concentrations suitable for nitrogen fixation was shifted to higher concentrations in the presence of carotenoids. The opposite effect was observed, when the carotenoidfree mutant W6 was compared with the parent, red coloured strain Sp Cd (not shown). In conclusion, carotenoids can exert some oxygen protection to Azospirillum and are important for nitrogen fixation and growth under certain oxygen stress conditions. It was possible to isolate carotenoid overproducing mutants from colourless wild type strains. We have preliminary evidence, that such mutants can be isolated from other wild type strains too.

2. MSX-resistant mutants

Nitrogen fixation of Azospirillum is repressed by reduced nitrogen compounds. This nitrogen control is mediated by the glutamine synthetase (GS) (6). Glutamine auxotrophic mutants of Azospirillum, defective in GS, are able to fix nitrogen in the presence of ammonium (7). We used the glutamine analogue methionine sulfoximine (MSX), a potent inhibitor of GS, to select for prototrophic, MSX resistant mutants with altered regulatory properties of GS. After mutagenesis, we isolated mutants of A. brasilense Sp Cd and A. lipoferum Sp RG 20a resistant to MSX (10 µg/ml) on minimal medium plates with and without (Sp RG 20a) ammonium (8). 137 mutants were tested in semisolid minimal medium with 5 mM initial ammonium concentration. While A. brasilense Sp Cd showed acetylene reduction only after a long lag phase, the mutant MSX 25 was active after a short lag (fig. 3). Another mutant with a similar behaviour was found among the MSX-resistant mutants of the A. lipoferum Sp RG 20a. When the ammonium concentration was determined in the growth pellicle during derepression, we could demonstrate higher nitrogen fixation rates at the same extracellular ammonium concentrations in the mutant MSX 25 as compared to the wild type (fig. 4). Furthermore, the mutant could grow to a higher stationary cell density than the wild type in semisolid medium with 5 mM initial ammonium supply.

Enzymatic analysis of the ammonium assimilation revealed, that the glutamate synthase activity was not altered in the mutant. However, the Mg-dependent GS activity and the ratio of Mg/Mn-dependent GS activity, which reflects the activity status of GS, were increased in the mutant in the derepression phase as compared to the wild type (see fig. 3). When nitrogen fixing cultures were treated with MSX (50 µM) for 60 minutes, the Mg-dependent GS activity in crude extracts was almost resistant to the MSX treatment (75 % of control), while the wild type was

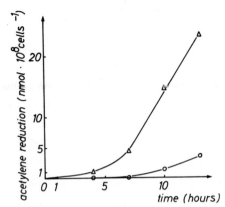

Figure 3: Derepression of nitrogen fixation of A. brasilense Sp Cd (O) and the mutant MSX 25 (△). Semisolid minimal medium, supplemented with 5 mM ammonium was inoculated with $2 \cdot 10^6$ bacteria per ml. After 13 hours incubation at 30° C (stagnant conditions) acetylene reduction was measured.

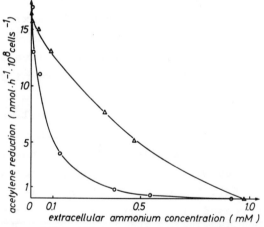

Figure 4: Nitrogen fixation activity of A. brasilense Sp Cd (O) and the mutant MSX 25 (△) at different extracellular ammonium concentrations. During derepression of nitrogen fixation in semisolid medium with 5 mM initial ammonium concentrations (see fig. 3) acetylene reduction and ammonium levels in the growth pellicle were determined.

fairly sensitive (35 % of control). In addition, the mutant GS was partially resistant to MSX in vitro. Therefore we suggest, that the mutant MSX 25 contains an altered glutamine synthetase. The mutation may influence the adenylation site of GS, leading to an inhibition of adenylylation. A conformational effect of MSX binding on the adenylylation site was recently demonstrated for the GS of E. coli (9). In conclusion, among MSX resistant isolates of Azospirillum, mutants with altered nitrogen regulation could be found. Since some of them are prototrophic and reveal higher nitrogen fixation and better growth yields at medium nitrogen levels, they may be useful to increase nitrogen supply of plants by Azospirillum.

3. Effects of an auxin overproducing mutant on plant root development

Recently the isolation and characterization of auxin overproducing mutants of A. brasilense Sp Cd were described in detail (1). Here we report on the influence of the auxin overproducing mutant FT 326 on root and shoot development of young maize plants in sterile quartz sand at different nitrate levels. As controls we examined pots, inoculated with rhizosphere microorganisms of maize, with the A. brasilense Sp Cd wild type and without inoculation. With 15 mmol l^{-1} nitrate in the soil we observed a decrease of root and shoot growth of one week old maize plants upon inoculation with A. brasilense Cd; the inhibition appeared to be more pronounced with the auxin overproducing mutant FT 326 (table 2 and fig. 5a). The diameter of the main roots were significantly increased in the presence of the mutant (table 2). Since the auxin production of the mutant was best at high nitrogen nutrition (1), the inhibition effects may be due to elevated, suboptimal levels of auxin in the rhizosphere. At low nitrogen concentrations (0.5 mmol l^{-1}), root length and diameter were increased upon inoculation with the mutant as with the Sp Cd wild type strain (table 2 and fig. 5b, c). While rhizosphere microorganisms inhibited root hair development, plants inoculated with A. brasilense Sp Cd and the mutant FT 326 showed a

Table 2: Influence of rhizosphere bacteria on the development of young maize plants in an axenic quartz system

	15 mmol l^{-1} nitrate in the soil[+]			0.5 mmol l^{-1} nitrate in the soil[++]		
	with maize rhizosphere bacteria	with A. brasilense Sp Cd	with the mutant FT 326	with maize rhizosphere bacteria	with A. brasilense Sp Cd	with the mutant FT 326
shoot dry weight	0.84 ± 0.1	0.89 ± 0.10	0.77 ± 0.10	0.66	0.94	0.98
number of roots	1.1 ± 0.3	0.60 ± 0.06	0.45 ± 0.12	1.32	1.22	0.97
total length of roots	0.94 ± 0.2	0.57 ± 0.03	0.57 ± 0.21	1.21	1.26	1.04
total length of main roots	0.68 ± 0.3	0.89 ± 0.16	0.67 ± 0.10	0.88	1.52	1.53
total length of lateral roots	1.06 ± 0.2	0.48 ± 0.03	0.52 ± 0.25	1.31	1.26	0.98
diameter of roots:						
main roots	0.89 ± 0.07	0.95 ± 0.06	1.09 ± 0.02	0.97	1.04	1.14
lateral roots	0.92 ± 0.08	1.02 ± 0.07	1.14 ± 0.07	1.06	1.12	1.23

The values in different lanes are normalized with the sterile controls, taken as 1.
[+] one week old maize plants, [++] two weeks old maize plants

Figure 5: Root and shoot growth of young maize plants under green house conditions in an axenic quartz sand system

a) Shoot and root growth of one week old maize plants inoculated with the mutant FT 326 at 15 mmol l^{-1} (left) and 0.5 mmol l^{-1} (right) nitrate in the soil.

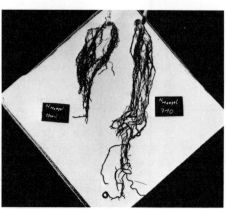

b) Seminal root development of two week old maize plants at 0.5 mmol l^{-1} nitrate without bacteria (left) and inoculated with <u>A. brasilense</u> Sp Cd (right)

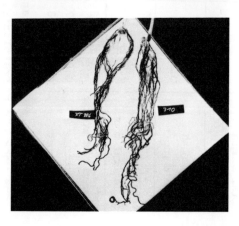

c) Seminal root development of two week old maize plants at 0.5 mmol l^{-1} nitrate in soil, inoculated with the mutant FT 326 (left) and the Sp Cd wild type (right).

root hair devolopment similar to the sterile control. Preliminary experiments with young rice plants at low nitrogen levels revealed a significant stimulation of root hair development in the presence of the auxin overproducing mutant. Further experiments at different developmental stages of the plant and with other plant species are needed to evaluate, if auxin overproducing strains could have beneficial effects on plant growth.

Acknowledgements

This work was supported by the Deutsche Forschungsgemeinschaft. We greatly appreciate the skilful technical assistance of Mrs. M. Ohlraun, Ms. Ch. Schemel and Ms. B. Sukacz.

References

1. Hartmann, A., Singh, M. and Klingmüller, W. (1983). Can. J. Microbiol. 69, in press.

2. Albrecht, S.L. and Okon, Y. (1980). In San Pietro, A., (ed.): Methods in enzymology, 69, pp 740-749. Academic Press, New York.

3. Nur, I., Steinitz, Y.L., Okon, Y. and Henis, Y. (1981). J. Gen. Microbiol. 122, 27-32.

4. Fawcett, J.K. and Scott, J.E. (1960). J. Clin. Path. 13, 156-159.

5. Kleinschmidt, J.A. and Kleiner, D. (1978). Eur. J. Biochem. 89, 51-60.

6. Magasanik, B. (1982). Ann. Rev. Genet. 16, 135-168.

7. Gauthier, D. and Elmerich, C. (1977). FEMS Lett. 2, 101-104.

8. Hartmann, A. (1982). In Klingmüller, W. (ed.): Azospirillum, Genetics Physiology Ecology. EXS 42, pp 59-68, Birkhäuser, Basel.

9. Shrake, A., Withley, Jr., E.J. and Ginsburg, A. (1980). J. Biol. Chem. 255, 581-589.

MOTILITY CHANGES IN AZOSPIRILLUM LIPOFERUM

T. HEULIN, P. WEINHARD and J. BALANDREAU

Centre de Pédologie Biologique, C.N.R.S.
B.P. 5, 54501 Vandoeuvre-Les-Nancy Cédix, France

Introduction

Motility is one of the most important taxonomic characters of the genus Azospirillum (10). However one of the Azospirillum isolated from the rhizosphere of rice in our laboratory, is non-motile (4, 11). We have established that this non motile strain (4T) belongs to the genus Azospirillum, using DNA-$_r$RNA hybridization (1). Moreover another A. lipoferum strain of the same origin, strain 4B (11), produced consistantly colonies of non motile individuals. This lead us to conduct a study of motility changes in strain 4B, and a few other Azospirillum strains (Table 1), as well as a comparative study of motile and non motile clones of strain 4B.

Table 1. Bacterial strains

Strain	Origin	Plant	Isolated by
A. brasilense			
Sp7 (ATCC 29145)	Brazil	Digitaria	Döbereiner (10)
R07	Senegal	Rice	Rinaudo (7)
51e	Brazil	Wheat	Döbereiner (10)
A95	France	Rice	This laboratory
A. lipoferum			
Br10	Brazil	Soil	Döbereiner (10)
B7C	France	Maize	This laboratory (6)
4B	France	Rice	This laboratory (11)
4T	France	Rice	This laboratory (11)

Frequency of motility variations

To eliminate the hypothesis, that the 4B strain was in fact a mixture of a classical Azospirillum and a non motile bacterium, a spontaneous mutant resistant to rifampicin (Rif^R) of the parental strain was isolated by plating on nutrient agar containing 40 µg.ml^{-1} rifampicin ; this mutant is thereafter refered to as 4B 40 N1.

Like the parental strain, 4B 40 N1 is able to produce colonies of non motile bacteria (Table 2) ; on nutrient agar these colonies are smaller than colonies of motile bacteria. The frequency of this variations is 0.85x10^{-3} per bacterium per generation (Table 3). Among several small colonies, the non motile 4B pHT2 clone was selected for further studies ; it was compared with strain 4B for carbohydrate metabolism, nitrogenase activity and optimal partial pressure of oxygen for growth.

Table 2. % of non motile bacteria for the 4B and 4B 40 N1 strains of A. lipoferum

Strain	Mean diameter of the colonies (mm) (72 hours at 28°C)	Number of colonies	Motility	% of non motile clones
4B	⩾ 3	154	+	4.3x10^{-2}
	< 1	7	-	
4B 40N1	⩾ 3	487	+	5.8x10^{-2}
	< 1.5	30	-	

Table 3. Frequency of the variation motile → non motile for the 4B 40 N1 strain of A. lipoferum. Frequency was evaluated by plating the content of 2 individual large colonies on nutrient agar and counting the small colonies obtained.
* Per bacterium per generation.

	Colonies with diameter > 3 mm = A	Colonies with diameter < 1 mm = B	Rate B/A	Initial bacteria in the colony	Number of generations	Frequency of the variations (*)
Colony 1	324	7	2.1x10^{-2}	6.6x10^6	22.7	0.9x10^{-3}
Colony 2	389	7	1.8x10^{-2}	10.0x10^6	23.3	0.8x10^{-3}

Acidification of carbohydrates

4B pHT2 and 4B were tested for the acidification of some carbohydrates with microtiter plates (1). All the sugars used had a final concentration of 10 g.l^{-1} and the pH of the medium which contained ammonium sulfate and bromocresol purple was ajusted to 7.1 with 1N KOH (10). The microtiter

Figure 1. Influence of the partial pressure of oxygen (pO$_2$) on the growth of 4B and 4B pHT2 (carbon source : glucose).

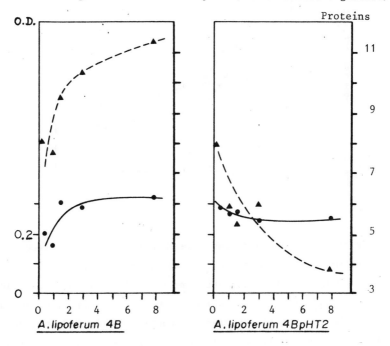

▲----▲ µg of bacterial proteins.ml^{-1} ; ●——● O.D. at 570 nm

Nitrogenase activity

Two different experiments were made to compare the motile and non-motile strains for their nitrogenase activity. First, a minimal N-free medium (3) where the carbone source was malate (10 mM) was inoculated with the two strains and incubated under 0, 1.0, 1.5 and 3.0 % of oxygen at 28°C. Nitrogenase activity was estimated by using the acetylene reduction method with incubation of the tubes under 10 % of C$_2$H$_2$. The optimum of nitrogenase activity was equivalent for the two strains (about 1 %) and this result confirms previous reports (3, 12). The specific activity is higher in the non motile

4B pHT2 strain than in the motile 4B strain (Figure 2). This difference between the two strains was further amplified in the second experiment.

Figure 2. Specific nitrogenase activity of 4B and 4B pHT2 under different concentrations of oxygen (carbon source : malate).

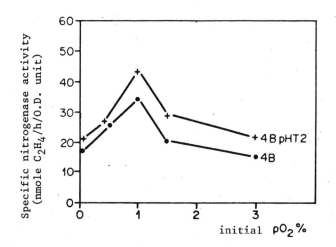

In this second experiment, the carbon source was a seedling of rice growing under monoxenic conditions in a spermosphere model (11). The bacteria were inoculated in a semi-solid N-free medium (13). The tubes were incubated under 10 % of C_2H_2. Ethylene evolution was measured after 2, 4, 10 and 17 days to determine the rate of the ethylene production (nmole C_2H_4 per day per seedling). The numeration of bacteria at the end of the experiment allowed the calculation of specific activity of nitrogenase (Table 5). Whereas the rate of ethylene production is much higher for the motile 4B strain, the non motile 4B pHT2 strain possesses the highest specific activity, mainly due to its weak growth (20.10^6 bacteria per tube instead of 18.10^7 for the motile strain). This result corroborates the first one.

Table 5. Specific nitrogenase activities of 4B and 4B pHT2 in spermosphere model (semi-solid N-free medium in which a rice seed germinates in the dark).

Strain	Rate of ethylene production (nmole C_2H_4/day/seedling)	Number of bacteria (per seedling)	Specific nitrogenase activity (nmole C_2H_4/day/10^6 bacteria)
A. lipoferum			
4B	1 670	18×10^7	9.3
4B pHT2	320	20×10^6	16.0

Reversion

The reversion of 4B pHT2 to some kind of motility was observed on the gelatin medium of STOCKER et al. (9). Motile individuals form larger colonies than the regular non motile 4B pHT2. The frequency of reversion (Table 6) seemed to be slightly higher than the frequency of appearance of non motile bacteria in 4B (1.8×10^{-3} per bacterium per generation).

Table 6. Frequency of the variation non motile → motile for the 4B pHT2 strain of A. lipoferum (for method see table 3) * per bacterium per generation.

	Colonies with diameter < 2 mm = A	Colonies with diameter > 10 mm = B	Rate B/A	Initial bacteria in the colony	Number of generations	Frequency of the variations (*)
Colony 1	82	2	2.4×10^{-2}	2.1×10^7	24.3	1.0×10^{-3}
Colony 2	91	6	6.6×10^{-2}	2.4×10^7	24.5	2.7×10^{-3}

Occurence in the genus Azospirillum

As a result of a preliminary survey of a few strains of Azospirillum it appears that similar changes in motility can be observed in other strains of A. lipoferum : we could evidence this in strains Br10 and B7C (Table 1). We also looked for changes in motility in four A. brasilense strains, Sp7 (ATCC 29 145), R07, 51e and A 95. In none of them could such variations be evidenced.

Conclusion

Regarding frequencies, motility changes reported here are not unlike the phase variation observed by SILVERBLATT (8) on Proteus mirabilis,

ZIED et al. (14) on *Salmonella typhimurium* and EISENSTEIN (2) on *E. coli* K12 CSH50. Other physiological traits associated with this phase variation along with its ecological significance and its genetic basis are currently under investigation.

References

1. Bally, R., Thomas-Bauzon, D., Heulin, T., Balandreau, J., Richard, C. and De Ley, 1983. Can. J. Microbiol. (in press).
2. Eisenstein, B.I. 1981. Science, 214, 337-339.
3. Franche, C. and Elmerich, C. 1981. Ann. Microbiol., 132A (1), 3-18.
4. Heulin, T., Bally, R. and Balandreau, J. 1982. Experientia Suppl., 42, 92-99.
5. Lowry, O.H., Rosenbrough, N.J., Farr, A.L. and Randall, R.J. 1951. J. Biol. Chem., 193, 265-275.
6. Mandimba, G. 1982. Thèse de Docteur-Ingénieur I.N.P.L., Nancy
7. Rinaudo, G. 1982. Thèse de Doctorat d'Etat, Université de Paris-Sud.
8. Silverblatt, F.J. 1974. J. Exp. Med., 140, 1696-1711.
9. Stocker, B.A.D., Zinder, N.D. and Lederberg, J. 1953. J. Gen. Microbiol., 24, 967-980.
10. Tarrand, J.J., Krieg, N.R. and Döbereiner, J. 1978. Can. J. Microbiol., 24, 967-980.
11. Thomas Bauzon, D., Weinhard, P., Villecourt, P. and Balandreau, J. 1982. Can. J. Microbiol., 28, 922-928.
12. Volpon, A.G.T., De Polli, H. and Döbereiner, J. 1981. Arch. Microbiol., 128, 371-375.
13. Watanabe, I. and Barraquio, W.L. 1979. Nature, London, 277, 565-566.
14. Zieg, J., Silvermann, M., Hilmen, M. and Simon, M. 1977. Science, 196, 170-172.

ATTRACTION OF AZOSPIRILLUM LIPOFERUM BY MEDIA FROM WHEAT-
AZOSPIRILLUM ASSOCIATION

D. Heinrich and D.Heß
Lehrstuhl für Botanische Entwicklungsphysiologie
Universität Hohenheim, Emil Wolffstr. 25,
7000 Stuttgart 70, FRG

Introduction

It has been observed that a wheat plant causes accumulation of Azospirilla in the rhizosphere of the plant. Many more bacteria accumulate on the root surface than in the soil nearby. The surface of the root is covered with a matrix of bacteria. This indicates that chemotaxis might play a role in the accumulation of bacteria in the rhizosphere.

Materials and Methods

The bacteria used in theses investigations was the strain Azospirillum lipoferum sp108$^{+)}$. Cultivation of the wheat: sterilization and preculture as described by Heß and Kiefer (1). For the chemotactic assay the capillary method of Currier and Strobel (2) and C^{14} radioactive labelled bacteria were used. The chemotaxis was measured by counting the radioactivity in the labelled bacteria that swam into the capillaries which were filled with the attractant. The chemotactic ability of eight different sugars (glucose, maltose, saccharose, fructose, arabinose, mannit galactose and xylose) and six different amino acids (methionin, serin, glutamin, asparagin, alanin, threonin) were tested. The chemotactic effect of wheat lectin on Azospirilla was tested too. Further more plant-media (PM), plant-bacteria-media (PBM), bacteria-media (BM), media alone (WH) and buffer as control (PBS) were examined in order to chemotactic attraction towards Azospirilla. These media were assayed from transfilter wheat-bacteria associations, established in our laboratories (Pudil and Heß (3)). Five replicates were used for each sample solution and at least five repetitions of each experiment were performed. The avrages and the confidence levels were calculated by the Student's "t" test.

Results and Discussion

The chemotactic effect of eight different sugars were tested. Azospirilla showed the highest chemotaxis towards saccharose (Fig.1). The lowest chemotactic ability was demonstrated by xylose and mannit, but both of them had a significant much higher attraction towards Azospirilla than buffer (PBS). Maltose, arabinose and galactose don't differ very much in their attraction ability but they attract distinct less than saccharose and distinct more than xylose and mannit. Glucose and fructose are located in the attractant range between xylose and mannit on the one side and the other sugars tested on the other side.

Fig.1: Chemotactic attraction of A. lipoferum sp108 by sugars. The tested sugars were dissolved in buffer (PBS) at a concentration of 10^{-3} M.

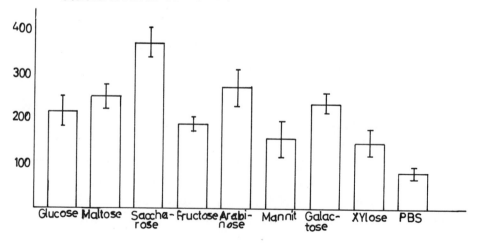

Amino acids do attract Azospirilla too (Fig.). The examination of six amino acids (methionin, serin, glutamin, asparagin, alanin and threonin) revealed glutamin as the amino acid with the most attractive effect followed by serin, methionin, alanin, asparagin and threonin at the end of the attractant range. All amino acids tested showed a significant much higher chemotactic ability than PBS.

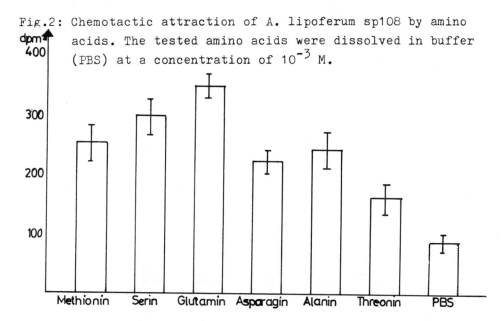

Fig.2: Chemotactic attraction of A. lipoferum sp108 by amino acids. The tested amino acids were dissolved in buffer (PBS) at a concentration of 10^{-3} M.

No chemotaxis could be demonstrated towards wheat lectin.
In our further chemotactic investigations we used different media assayed from transfilter associations between Azospirilla and wheat plants. The results are demonstrated in Fig.3. Bacteria were attracted most by PM (plant-media). The chemotactic attraction by PBM - the media in which an effective association between bacteria and wheat was established - was significant lower. An explication of this lower attraction by PBM could be that a part of the attractants were used up in the association. Both PM and PBM do attract Azospirilla in a much higher rate Than BM (bacteria-media) and media alone (WH). Between BM and WH there was no significant difference in attraction of Azospi-

rilla. Whereas BM and WH were significant more attractive than
PBS. This is not astonishing if we have a look to the composition of WH (Wagner and Heß (4)).

Fig.3: Chemotactic attraction of A. lipoferum sp108 by media
from wheat-Azospirilla associations: PM (plant-media),
PBM (plant-bacteria-media), BM (bacteria-media), WH
(media alone) and as control PBS (buffer).

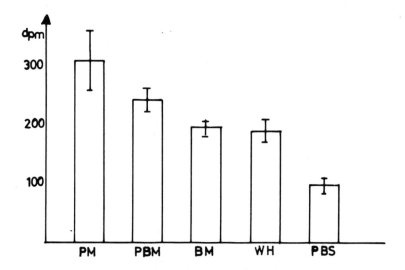

A concentration range of PM was chemotactically tested too.
The chemotactic attraction of PM 2-fold concentrated was significant much higher than the attractive effect of PM 1-fold and
this for its part was significant higher than that from PM 1:2
(one part PM and two parts PBS). Between PM 1:2 and PM 1:5 (one
part PM and five parts PBS) there was no significant difference
and both of them showed no significant higher chemotactic attraction in comparison to WH. These results confirm the supposition
that in the wheat root exudates there are substances which
attract Azospirilla. In Pm 1:2 and in PM 1:5 these substances
were diluted so much that they have no effect on Azospirilla
any longer.

Our experiments revealed, nevertheless, that Azospirilla respond chemotactically to different sugars, amino acids and root exudates from wheat plants. This indicates that the wheat root exudates contain substances which attract chemotactically Azospirilla . This is a hint that one of the first steps in establishing associations between Azospirilla and wheat probably is initiated by the chemotactic attraction of Azospirilla towards wheat exudates. Detailed investigations of the nature of the attractants in the wheat root exudates towards Azospirilla is chemotactic are in work.

References

1. Heß, D. and Kiefer, S. 1981, Z.Pflanzenphys. 101, 15-24.
2. Currier, W.W. and Strobel, G.A. 1976, Plant Physiol. 57, 820-823.
3. Pudil, H. and Heß, D. 1983, in press.
4. Wagner, G. and Heß, D. 1973, Z. Pflanzenphys. 69, 262-269.

+) see De Polli et al. in Vose, P.B. and Ruschel, A.P. eds.: Associative Nitrogen Fixation. CRC Press Inc., Boca Raton, Florida, 1979.

NITROGEN FIXATION AND DENITRIFICATION BY A WHEAT-AZOSPIRILLUM ASSOCIATION

H.BOTHE, A.KRONENBERG, M.P.STEPHAN, W.ZIMMER and G.NEUER

Botanisches Institut, Universität Köln, Gyrhofstr. 15, D-5 Köln 41

Abstract

An association between wheat and Azospirillum brasilense Sp.7 has been studied for N_2-fixation (C_2H_2-reduction) and denitrification activity. The association was grown in semi-solid agar medium for a week in air and then assayed under microaerobic conditions. When the amount of nitrate was less than 1 mM in the medium, the association commenced to perform C_2H_2-reduction 3-5 h after the removal of air. C_2H_4-formation stopped above 1 mM nitrate in the medium, and the association evolved N_2O by denitrification. C_2H_2-reduction and N_2O-formation were strictly dependent on the presence of both Azospirillum and wheat plants in the assays. The addition of carbohydrates to the medium did not enhance each of the activities indicating that the bacteria must have lived from carbon sources of the plants. C_2H_2-reduction and N_2O-formation were marginal at a temperature prevailing in the temperate climate zone and maximal at a temperature comparable to those of tropical regions. Some C_2H_2-reduction but no N_2O-formation activities were observed when the assays were performed in air.

Introduction

The association between roots of grasses and Azospirillum species has attracted special attention in the past few years. N_2-fixing Azospirillum was suggested to provide fixed nitrogen to crop plants which could increase crop productivity or reduce the requirements for fixed nitrogen fertilizers or both. Results of greenhouse or field experiments have not yet been conclusive. The majority of the reports, however, gave positive indications (for reviews see 1,2). In addition to supplying the plants with fixed nitrogen, Azospirillum may augment plant growth by excreting phytohormones such as auxins, cytokinins and gibberellins (3-5) or by stimulating the uptake of $NO_3^-, K^+, H_2PO_4^-$ or other mineral ions into plants (6). The issue may be disturbed by the fact that Azospirillum can also perform denitrification (7-9). Recent experiments have shown that Azospirillum can even grow with nitrate as the terminal respiratory electron acceptor in batch culture(10). When the O_2-concentration is low in the culture, nitrate is converted to nitrite and partly to N_2O and N_2. Up till-date, denitrification has only been studied in pure cultures of Azospirillum and not in the grass-Azospirillum association. It is conceivable that crop plants may lose combined nitrogen by the denitrification capability of Azospirillum when the levels of nitrate are high and those of O_2 are low in soils.

In an attempt to understand the interactions between crop plants and Azospirillum, we have started to do experiments under defined laboratory conditions using wheat and the type strain, Azospirillum brasilense Sp.7. The present communication will indicate that the Azospirillum-wheat association either performs N_2-fixation (C_2H_2-reduction) or denitrification, depending on the amount of oxygen and nitrate available

Materials and Methods

For the experiments with the wheat-Azospirillum association, Azospirillum brasilense Sp.7 was grown as batch culture in an autoclaved medium containing in g/l: $MgSO_4 \cdot 7 H_2O$, 0.2; NaCl, 0.1; $CaCl_2 \cdot 2 H_2O$, 0.02; $NaMoO_4 \cdot 2 H_2O$, 0.02; $MnSO_4 \cdot H_2O$, 0.01; KH_2PO_4, 0.25; $FeSO_4 \cdot 7 H_2O$, 0.0069; ethylenediaminetetraacetic acid, 0.0093; DL-malate, 5. $FeSO_4$-ethylenediaminetetraacetic acid and phosphates were autoclaved separately and added to the medium after cooling. The pH was adjusted to 6.9 prior to autoclavation. After growth for 24 h at $30^{\circ}C$, 1 ml of the bacterial suspension containing approximately 1.3×10^9 cells was used as the inoculum.

Wheat (Triticum aestivum L. Ralle) seeds were disinfected by treatment with 70 % ethanol for 10 min, followed by treatment with

0.1 % $HgCl_2$ dissolved in 0.05 N HCl for 10 min and by washing 5 x with sterilized destilled water. The seeds were placed on sterile and wet filter papers in Petri dishes and germinated for 2-3 d in a growth chamber with light/dark cycles (12.5 h in the light at $33^\circ C$, 11.5 h in the dark at $23^\circ C$).

The experiments were performed in 1 l flasks containing 100 ml sterilized medium with 0.8 % agar (Merck) and the following salts in g/l: $CaCl_2 \cdot H_2O$, 0.14; $MgSO_4 \cdot 7 H_2O$, 0.17; $MnSO_4 \cdot H_2O$, 0.01; KH_2PO_4, 0.14; $Na_2HPO_4 \cdot 12 H_2O$, 0.21; $FeSO_4 \cdot 7 H_2O$; 0.04 and KNO_3 as indicated in the legends of the figures and the tables. 25 germinated wheat seeds and 1 ml of the bacterial suspensions were given onto the agar surface of the flasks under sterile conditions. The flasks were sealed with sterile cotton wool and incubated for 7 d in the growth chamber at light/dark cycles (12.5 h in the light at $33^\circ C$ and 11.5 h in the dark at $23^\circ C$). The cotton wool was then replaced by sterilized rubber stoppers, and the flasks were then repeatedly evacuated and gassed with argon to remove the air. The O_2-content in the gas phase was approximately 0.2 % at the beginning of the experiment. After injection of 5 ml C_2H_2, the flasks were incubated for 24 h in the growth chamber under the conditions just mentioned. C_2H_4-formation was determined by gas chromatography using a flame ionization detector and a Porapak R column, and N_2O and N_2 in a gas chromatograph fitted with a thermal conductivity detector and a Porapak Q column with He as the carrier gas.

For the experiments of fig. 1 and 2 (pH curves for NO_2^- or N_2O utilization by Azospirillum in liquid culture), Azospirillum brasilense Sp. 7 was grown in continuous culture with nitrite as terminal respiratory electron acceptor. Details of these growth conditions will be published elsewhere(M. P. Stephan, W. Zimmer, H. Bothe, in preparation). Azospirillum taken from the fermenter was centrifuged (10 min at 12 000 x g at room temperature) to concentrate the cells 2 - 3 fold and to remove NO_2^-. The pellet was then suspended in the growth medium in which combined nitrogen (NO_3^- or NO_2^-) was omitted. The cells were assayed in 7.2 ml Fernbach flasks containing in a final volume of 3 ml: 100 mM TES or

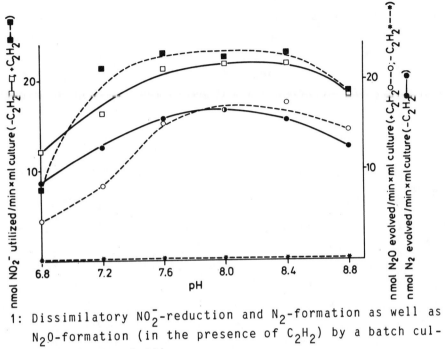

Fig. 1: Dissimilatory NO_2^--reduction and N_2-formation as well as N_2O-formation (in the presence of C_2H_2) by a batch culture of _Azospirillum_

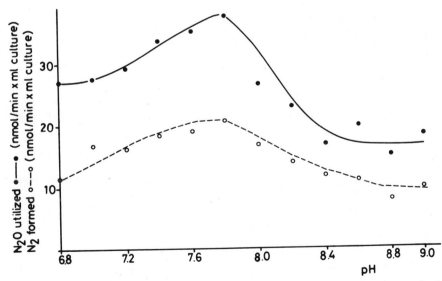

Fig. 2: Dissimilatory N_2O-utilization and N_2-formation by a batch culture of _Azospirillum_

Tricine-buffer, pH as indicated in fig. 1 and 2, 6 mM $NaNO_2$ or 9 mM N_2O (in the gas phase), 2.2 ml cells (optical density at 560_{nm}=0.43) and 7 mM C_2H_2 when indicated. The reaction was performed at 30°C for 4 h under argon in a shaking water bath, and gas utilization or formation was determined by gas chromatography.

Results

Azospirillum brasilense Sp. 7 grows in the absence of O_2 using NO_3^- as terminal respiratory electron acceptor (10). The present communication shows that the bacterium also utilizes NO_2^- (fig.1) or N_2O (fig.2) when neither NO_3^- nor O_2 are available. In respiratory NO_2^--reduction, the stoichiometry between NO_2^--disappearance and N_2-formation is approximately 2:1, and neither free N_2O (fig. 1) nor ammonia (not documented) are formed. Nitrite is, however, quantitatively converted to N_2O in the presence of C_2H_2 in the assay vessels (fig. 1). C_2H_2 specifically blocks N_2O-reductase among the enzymes of dissimilatory nitrate reduction (see 11). The stoichiometry between N_2-formation (in the absence of C_2H_2) and N_2O-production (in the presence of C_2H_2) is approximately 1:1. NO_2^--utilization and N_2-formation as well as N_2O-production (in the presence of C_2H_2) are independent of the H^+-concentration in the assays between 7.6 and 8.4 (fig. 1). Fig. 2 shows the utilization of N_2O as respiratory electron acceptor by *Azospirillum*. N_2O is mainly converted to N_2. As the stoichiometry between N_2O-consumption and N_2-production does not approach to unity, N_2 may partly be reduced further to ammonia which is, however, not excreted by the bacterium under these assay conditions. N_2O-utilization and N_2-formation show an optimum at pH 7.8.

The knowledge of these experiments with liquid cultures of *Azospirillum* served as the background for the tests with the wheat-*Azospirillum* association. For these experiments, germinated wheat seeds and *Azospirillum* were placed onto the surface of a semisolid agar medium and incubated for a week in a growth chamber under light/dark cycles. After this, the C_2H_2-reduction and N_2O-formation activities of the association were assayed for 24 h un-

Table 1: Assay conditions for C_2H_4- and N_2O-formations by the wheat-*Azospirillum* association

Assay condition	C_2H_2-reduction		N_2O-formation	
	$-KNO_3$	$+KNO_3$	$-KNO_3$	$+KNO_3$
1. complete (=+ *Azospirillum*, + wheat)	5.8	0.4	0.0	9.1
2. - *Azospirillum*	0.04	0.04	0.0	0.0
3. - wheat	0.02	0.02	0.0	0.0
4. wheat, where shoots were excised at the inoculation	0.9	0.3	0.0	4.5
5. complete + malate (1 mM)	7.8	1.4	0.0	7.4
6. complete + arabinose (1 mM)	4.8	0.2	0.0	8.3
7. complete + sucrose (1 mM)	6.3	1.4	0.0	8.0
8. complete + glucose (1 mM)	4.1	0.2	0.0	7.5

25 germinated wheat seeds and 1 ml *Azospirillum* culture containing approximately 1.3×10^9 cells were inoculated onto the surface of a semisolid agar/mineral salt medium in 1 l flasks and incubated for a week in a growth chamber (see Materials and Methods). C_2H_2-reduction and N_2O-formation activities were then assayed for a day where the gas phase consisted of argon supplemented with 0.5 % C_2H_2 and 1 % O_2. Rates are given in μmol C_2H_4 or N_2O formed / d x flask and standard deviations are 1-2 μmol/ d x flask for all measurements.

der a gas phase consisting of argon supplemented with 0.2 % O_2 and 0.5 % C_2H_2 (see Materials and Methods). Significant C_2H_2-reductions and N_2O-formations were reproducibly observed in the presence of both plants and Azospirillum in the flasks. Table 1 indicates that no C_2H_2-reduction and no N_2O-formation were detectable when either plants or Azospirillum were omitted. The addition of nitrate to the medium drastically reduced the daily C_2H_4-formation activity. On the other hand, N_2O-formation was only observed with associations grown in the presence of nitrate and assayed with C_2H_2 in the gas phase. C_2H_2-reduction and N_2O-formation were dependent on intact plants, because the activities decreased when the shoots were excised at the beginning of the incubation. C_2H_4-formation and N_2O-production did not occur at the expense of malate from the culture medium in which Azospirillum had

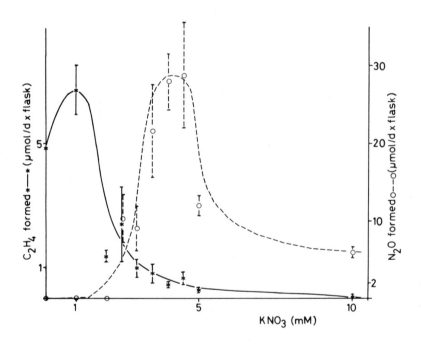

Fig. 3: N_2O- formation and C_2H_2-reduction by the wheat - Azospirillum association: Dependence on nitrate in the medium

been grown prior to inoculation. This statement can be taken from the finding that controls without wheat plants did not show any C_2H_2-reduction or N_2O-formation. The addition of malate, arabinose, sucrose or glucose did not enhance C_2H_4- or N_2O-formations (table 1). Wheat plants obviously supply enough carbohydrates to support both activities in Azospirillum. On the average, the rates of N_2O-formation were 1.5 - 3 fold higher than those of C_2H_4-production.

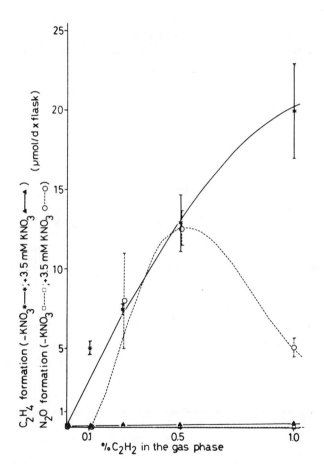

Fig. 4: N_2O-formation and C_2H_2-reduction of the wheat -Azospirillum association: Dependence on C_2H_2 in the gas phase of the vessels

The activities of N_2O-formation and C_2H_2-reduction of the wheat-<u>Azospirillum</u> association were dependent on the amount of nitrate in the agar medium (fig. 3). C_2H_2-reduction was drastically inhibited by the addition of more than 1 mM KNO_3 to the association. In contrast, N_2O-formation required the presence of more than 1 mM KNO_3 during growth of the association, and was optimal between 3-5 mM with higher concentrations being inhibitory (fig. 3). This decrease was due to unspecific effects, because the wheat plants showed impaired growth at such high KNO_3-concentrations. As expected, C_2H_2 was required for both C_2H_4- and N_2O-formations (fig. 4). C_2H_4-formation was therefore due to the plant-bacteria association and not to abiotic effects. The reactions were saturated at about 1 % C_2H_2 in the gas phase for C_2H_4-formation and at 0.5 % C_2H_2 for N_2O-formation. Higher concentrations than 0.5 % C_2H_2 in the gas phase decreased N_2O-formation for some unknown reason.

Fig. 5a and fig. 5b give the kinetics for C_2H_2-reduction and N_2O-formation, respectively. When the wheat - <u>Azospirillum</u> association, grown in the growth chamber for a week, was incubated under argon plus C_2H_2, both C_2H_2-reduction and N_2O-production commenced after a lag phase of 3 - 5 h and then proceeded linearly during the next 20 h. Microscopic examinations showed that the motile bacteria had moved from the agar surface to the roots. As already judged by examination with the eye, <u>Azospirillum</u> had multiplied in the neighbourhood of the roots during growth of the association for a week. We have, however, not been able to quantify growth of the bacteria, because it was difficult to separate them from the agar.

N_2O-formation strictly required the removal of air (table 2). C_2H_2-reduction per day and association was also considerably enhanced by replacing the air by argon. The O_2-content in the gas phase was low then (approximately 0.2 %). However, it must be kept in mind that part of the O_2 could probably not be removed from the roots sticking in the agar and that the plant leaves produced O_2 photosynthetically during the assay. Remarkably, some C_2H_4-formation was also detected under aerobic assay conditions (table 2).

Both C_2H_2-reduction and N_2O-formation activities were much dependent on the growth temperature of the association in the chamber (table 3). Activities were only marginal when the association was grown at 23° in the light and 16° C in the dark (conditions comparable to the temperate climate zone) and not as usual at 33° in the light and 23° in the dark (conditions of tropical regions).

Table 2: C_2H_2-reduction and N_2O-production by the wheat-<u>Azospirillum</u> association under aerobic and microaerobic conditions

assay condition	C_2H_2-reduction		N_2O-formation	
	$-KNO_3$	$+KNO_3$	$-KNO_3$	$+KNO_3$
under argon- 0.2 % O_2	6.9	0.9	0.0	12.6
in air	0.4	0.2	0.0	0.0

Data are given in μmol C_2H_4 or N_2O formed/d x flask. The experimental conditions are the same as in table 1.

Table 3: C_2H_2-reduction and N_2O-formation by the wheat-<u>Azospirillum</u> association: Comparison of the activities when the association was grown at higher and at lower temperature

assay condition	C_2H_2-reduction		N_2O-formation	
	$-KNO_3$	$+KNO_3$	$-KNO_3$	$+KNO_3$
1. 12.5h at 33° in the light 11.5h at 23° in the dark	4.9	0.8	0.0	21.6
2. 12.5h at 23° in the light 11.5h at 16° in the dark	0.3	0.1	0.0	2.2

Data are given in μmol C_2H_4 or N_2O formed/d and flask. The experimental conditions are the same as in table 1

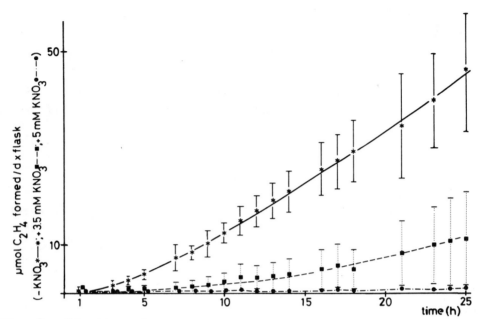

Fig. 5a: Kinetics of the C_2H_2-reduction by the wheat-<u>Azospirillum</u> association

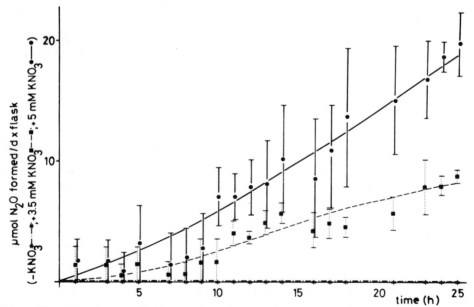

Fig. 5b: Kinetics of the N_2O-formation by the wheat-<u>Azospirillum</u> association

Discussion

The present communication shows that the wheat-<u>Azospirillum</u> association performs either C_2H_2-reduction or N_2O-formation, depending on the growth conditions in the semisolid agar medium. C_2H_2-reduction by the bacteria proceeds only at concentrations of nitrate smaller than 1 mM in the cultures. All bacteria preferentially meet their nitrogen requirements by assimilatory nitrate reduction and not by N_2-fixation, and <u>Azospirillum</u> is no exception to the general rule (10,12). In the present investigation, the addition of carbohydrates did not stimulate C_2H_2-reduction which means that the bacteria must have lived from the carbon compounds of the wheat plants. C_2H_2-reduction activity of the association was greatly enhanced by removing the air out off the flasks. It was pointed out by several authors that long incubation times at low O_2-concentrations lead to an overestimation of the nitrogenase activity occurring in situ(1,2). In the present investigation, C_2H_2-reduction commenced as early as 3-5 h after the removal of air. Some activity was also observed when the assays were performed in air. Little is known about the O_2-concentrations prevailing at the root surface or in the root cortex. It is conceivable that compartments in the roots exist where the O_2-levels are low due to respiration of the plant cells and where the bacteria find optimal conditions even in natural environments.

C_2H_2-reduction activity in the present investigation did not show diurnal fluctuations and was not lowered during the dark period as observed by others(13). Such differences in the results are probably due to different experimental conditions. In the present communication, it was extremely important to keep the conditions (number of plants and bacteria, incubation time, temperature etc.) to get reproducible results. Such experiments also require more repetitions than physiological or biochemical tests for statistically sound data. All our data are averages of at least 10 sets of experiments for every figure and table.

Remarkably, C_2H_2-reduction was greatly enhanced when the association was grown at higher temperature. Rather little activity was observed at conditions prevailing in temperate climate soils. Thus temperature appears to be particularly important for maximal nitrogenase activity of the association between wheat and Azospirillum brasilense Sp.7. Azospirillum strains are particularly abundant in tropical soils. The contributions of Azospirillum to the nitrogen economy in temperate soils are presumably modest, although positive evidence has been forwarded also for the association grown in cold-climate soils (13,14,15).

The present communication also shows that the wheat-Azospirillum association performs denitrification provided the NO_3^--concentration in the medium is high and the O_2-level in the gas phase is low. All denitrifying bacteria, including Azospirillum, preferentially utilize O_2 as respiratory electron acceptor. In the absence of O_2 or at low O_2-concentrations, Azospirillum brasilense Sp.7 dissimilates NO_3^- to N_2 via NO_2^- and N_2O. In the present investigation, the experiments were performed in the presence of C_2H_2, because nitrogenase and denitrification activities can be tested simultaneously in one and the same vessel and because N_2O-formation is more easily measured than N_2-production. When artificial plant-Azospirillum associations are to be constructed for applications, Azospirillum strains are available which have nitrogenase and which convert nitrate to nitrite but not to gaseous N_2 or N_2O. (8). It is conceivable that naturally occurring strains living in associations with grasses perform denitrification to a significant extent when the NO_3^--content is high and the O_2-level is low in soils (e.g. under waterlogged conditions).

In the present communication, the experiments were done with semisolid agar and in a mineral medium in order to simplify and to get well-defined experimental conditions. Growth of the wheat plants was not significantly enhanced by N_2-fixation or reduced by denitrification of the bacteria. Such enhancement or reduction was not expected to occur, since young wheat plants are sufficiently supplied with nitrogen from the protein reserves of the seeds. Further experiments have to show whether N_2-fixation can increase plant growth at later stages of the plant development and whether denitrification does the opposite.

References

1. Döbereiner,J. and De-Polli,H. 1980.In: Nitrogen fixation, Stewart,W.D.P. and Gallon,J.R. (eds.), pp. 301-333, Academic Press,London

2. Van Berkum,P. and Bohool,B.B. 1980. Microbiol. Rev. 44, 491-517

3. Umali-Garcia,M.,Hubbell,D.H., Gaskin,M.H. and Dazzo,F.B. 1980. Appl. Environm. Microbiol. 39,219-226

4. Martin,P. and Glatzle A. 1982.Experientia Suppl. 42, 108-120

5. Hartmann,A., Singh,M. and Klingmüller,W. 1983.Can. J. Microbiol. in press

6. Lin,W., Okon Y. and Hardy, R.W.F. 1983.Appl. Environm. Microbiol. 45, 1775-1779

7. Neyra,C.A. and van Berkum,P. 1977. Can.J.Microbiol. 23, 306-310

8. Magalhaes,L.M.S., Neyra,C.A. and Döbereiner,J. 1978. Arch. Microbiol. 117, 247-252

9. Scott,D.B.,Scott, C.A. and Döbereiner,J. 1979. Arch. Microbiol. 121,141-151

10. Bothe,H., Klein,B.,Stephan,M.P. and Döbereiner,J. 1981. Arch. Microbiol. 130, 96-100

11. Payne,W.J. 1981. In: Denitrification, nitrification and atmospheric nitrous oxide, Delwiche,C.C. (ed.),pp. 85-103,Wiley & Sons,New York.

12. Bothe,H., Barbosa,G. and Döbereiner,J. 1983.Z. Naturforsch. in press.

13. De-Polli,H.,Boyer,C.D. and Neyra,C.A. 1982. Plant Physiol. 70, 1609-1613

14. Haahtela,L., Wartiovaara,T. and Sundman,V. 1981. Appl. Environm. Microbiol. 41, 203-206

15. Haahtela,L., Kari,K. and Sundman,V. 1983. Appl. Environm. Microbiol. 45, 563-570

EFFECT OF OXYGEN CONCENTRATION ON ELECTRON TRANSPORT COMPONENTS AND MICRO-
AEROBIC PROPERTIES OF AZOSPIRILLUM BRASILENSE

Y. OKON, I. NUR and Y. HENIS
Department of Plant Pathology and Microbiology,
The Hebrew University of Jerusalem,
Faculty of Agriculture, Rehovot, 76100, Israel.

Introduction

Within the past two decades the electron transport chains of N_2- fixing bacteria have been studied and characterized (6,9,17). Some of the outstanding features of the respiratory chains of N_2-fixing bacteria are their extreme diversity, multiplicity in the arrangements of cytochromes and of quinones. In addition, many bacteria are capable of altering their cytochrome content in response to changes in growth conditions (7). Rhizobium and Azotobacter posses a large array of primary dehydrogenases, and most strains contain multiple terminal oxidases (1,4,6,8).

The physiology of Azospirillum brasilense strain Cd has been studied at constant dissolved O_2 concentrations (d.o.t.) in a chemostat (12). The organism readily adapted to different d.o.t. levels. Under high d.o.t. it produced carotenoids (12,13) and increased markedly succinate oxidase and NADH oxidase activities, whereas under low d.o.t. the above enzyme activities were reduced.

In this work the nature of some components of the respiratory electron transport chain, from low and high d.o.t. chemostat cultures of A. brasilense is reported.

Materials and Methods

Organisms and growth conditions

Cultures of Azospirillum brasilense strain Cd ATCC 29729 (11) were maintained on N-free malate medium (15). The bacteria were grown in a bench chemostat (New Brunswick-NBS-Model C-30) of 1.5 litre working capacity. Levels of dissolved oxygen tensions (d.o.t.) were maintained constant during growth by a sterile mixture of N_2 and air (12). The two levels of d.o.t.

used were 0.007 atm and 0.2 atm. The bacterium was grown at 30°C at a constant dilution rate (generation time 8.66h) in mineral growth medium containing potassium phosphate buffer pH 6.8 (15), supplemented with 2 g malic acid, 0.8 g NaOH and 0.5 g NH_4Cl per litre. The cultures were harvested by centrifugation for 10 min at 6000 x g and washed twice in one tenth of the original volume of 0.1M tris hydrochloride buffer pH 7.3 (N_2 saturated, oxygen free) containing 15 mM magnesium acetate. The bacteria was suspended in 4 to 5 ml of the same buffer and used immediately for either measurements of intact bacterial respiration or stored at -20°C for preparation of cell extracts.

Preparation of extracts and cell fractions

Cells were ruptured by ultrasonication (1 min intervals) 1.5A with an MSE ultrasonic disintegrator and centrifuged at 20,000 x g for 20 min. To recover a membrane fraction (5) the supernatant was centrifuged at 144,000 x g for 4 h.

Centrifugation of extracts from high d.o.t. cultures resulted in two layers. A membrane particle fraction and a lighter cloudy band over the first layer. The layers were separated by a sucrose gradient (12 ml centrifuge tubes combining a lower layer of 70% (w/v) sucrose and upper sucrose gradient 60-30%). The tubes were centrifuged at 60,000 x g under swing out conditions in a Sorval OTD-50 for 4 h.

For respiration experiments, membrane particles were suspended in 0.1 M Tris acetate/5 mM magnesium acetate buffer pH 7.2. The terminal oxidase activities were measured with a Clark-type electrode covered with ultra thin teflon membrane in a closed reaction vessel (1.5 ml) maintained at 30°C. This apparatus was sensitive enough to measure 0.1 µM of O_2. The effect of inhibitors on the respiration rate was evaluated by measuring (after injection of 100 µl oxygen saturated ice-cold suspension in buffer mixture: 44.3 nmols of oxygen) the time needed by the suspension to become anaerobic. The reaction mixture (1.5 ml final volume) contained approximately 1 mg protein, 75 µmol Tris-acetate buffer pH 7.3, 15 µmol magnesium acetate and either 10 µmol succinate or 2 µmol NADH. Antimycin A was added in 10 µl of ethanol. This amount of ethanol alone had no effect on the oxidase activity. The anaerobic reaction mixture was allowed to stand for 5 min in the presence of the inhibitor, before the oxygen was added to start the reaction. Cytochromes were examined spectroscopically with U.V. Ikon 820 in the split-beam made with a band pass of 1 nm. Difference spectra measurements were made with dithionite-reduced against ferricyanide-oxidized membrane preparations in anaerobic cuvette.

Proteins in cell-free extracts were determined by the modifications of Lowry's procedure (10) using bovine serum albumen (Sigma) as standard.

Results

Separation of membrane fractions from high d.o.t. cultures

Centrifugation of cell free extracts of A. brasilense grown under high d.o.t. resulted in two pellet layers; a heavy, red-brown membrane layer and a lighter, diffuse red layer with low weight membrane particles (8). The diffuse layer was not obtained from low d.o.t. cell free extracts. The membrane and low weight membrane particles were separated in a sucrose density gradient. By using this separation, the light particles were cleansed from other soluble proteins. In both membrane and light particle fractions there was succinic oxidase activity. The diffuse fraction contained two defined peaks, cytochrome-b 560 nm (α peak) and 535 (β peak) and cytochrome $a-a_3$ 603-605 nm (data not shown).

Respiratory chain components from high d.o.t. and low d.o.t. cultures

In both the cells that were grown under high or low d.o.t., there was an identical amount of soluble cytochrome c in the supernatant (about 1.08 µmol/mg protein, as calculated from wavelength pairs EmM = 17.3, 551-538 nm; Tissiers (1956) (16) for A. vinelandii cytochrome $c_4 + c_5$).

In both high or low d.o.t. membrane preparations, there were peaks at 551 nm, which could be composed of various cytochromes c with fused maxima at 551 nm. Cytochrome b (maxima of 428 nm and 560 nm) was tentatively identified in both high and low d.o.t. preparations and a cytochrome with a peak at 603 nm from high d.o.t. cells was tentatively identified as cytochrome $a-a_3$. Only traces of $a-a_3$ fraction were present in membranes of cells grown under low d.o.t.

Similar soluble and membrane bound cytochromes were found in Azotobacter and in Rhizobium (6,8,18). The $a-a_3$ fraction has been reported to be absent from R. japonicum bacteroids in soybean nodules, but largely distributed in cell cultured Rhizobium (1).

Respiratory inhibition of CO

Intact cells taken from high or low d.o.t. cultures showed different responses to CO inhibition. At a ratio of 2.5 pCO/pO_2, A. brasilense from air saturated cultures recovered only 12% of its original respiration rate. Cultures from oxygen limiting conditions recovered 106% of the original respiration activity. The same phenomenon was demonstrated in membrane preparations (Fig. 1). Carbon monoxide strongly inhibited the succinate oxidase of high d.o.t. membranes (Fig. 1), and NADH oxidase activity (data not shown). At a ratio of 2 pCO/pO_2 almost 75% of the original oxygen uptake was inhibited. No inhibition was found in membranes from low d.o.t. cultures. Furthermore, CO slightly enhanced the membrane respiration rate (Fig. 1). This fact supports the assumption that A. brasilense at low d.o.t. possess a terminal oxidase other than the cytochrome $a-a_3$ complex. It has been suggested (1) that in microaerobic Rhizobium bacteroids cytochrome b and c might be linked with terminal oxidases other than cytochrome $a-a_3$.

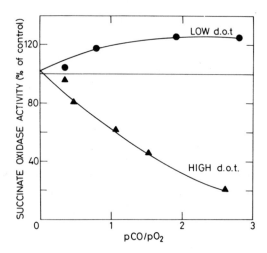

Fig. 1: Inhibition of succinate oxidase by CO in membrane fractions prepared from high (▲—▲) and low (●---●) d.o.t. A. brasilense-Cd cultures. 100% Activity was 145 μLO_2/h/mg protein of bacteria grown under low d.o.t. and 105 μLO_2/h/mg protein for bacteria grown in high d.o.t.

Respiratory inhibition by KCN on membrane fractions

The effect of KCN on NADH oxidase was compared with the effect on succinate oxidase activity in high d.o.t. membranes (Fig. 2a). When NADH was used as substrate, the activity was markedly inhibited by more than 40%, by 5 x 10^6M KCN (Fig. 2a). On the contrary, when succinate was used, succinate

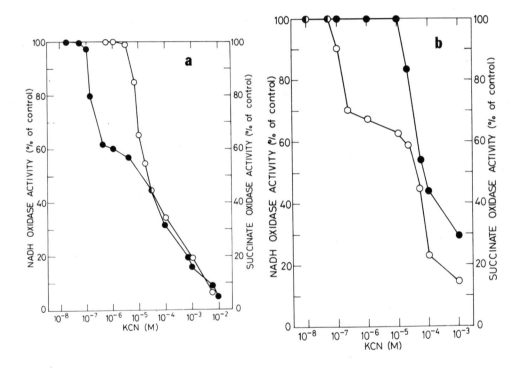

Fig. 2: Inhibition of NADH (●——●) and succinate oxidase (O-O) in membrane fraction by various concentrations of KCN. a) Membrane fraction prepared from A. brasilense-Cd and grown in high d.o.t. 100% activity was 136.3 µLO_2 for NADH oxidase and 88.8 µLO_2/h/mg protein for succinate oxidase. b) Membrane fraction prepared from A. brasilense-Cd grown in low d.o.t. 100% activity was 223 µLO_2/h/mg protein for succinate oxidase and 205 µLO_2/h/mg protein for NADH oxidase.

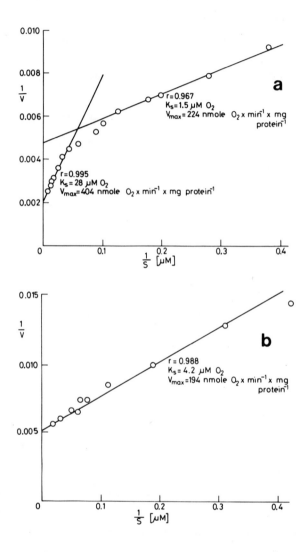

Fig. 3: Double reciprocal plots of respiration in high d.o.t. membrane particles of <u>Azospirillum brasilense</u>-Cd. a) Without KCN. b) With 1×10^{-4} M KCN. V is expressed as nmol O_2/min per mg protein and S as $\mu M O_2$.

oxidase was affected only by high concentrations of KCN (5×10^{-5} M KCN did not affect the activity at all; Fig. 2a). Furthermore, NADH oxidase showed a bisphasic pattern of inhibition which was not observed with succinate oxidase. This biphasic pattern strongly suggests the presence of a branched respiratory electron transport chain in high d.o.t. membranes of A. brasilense-Cd.

In the low d.o.t. membranes, the same pattern of inhibition was observed, but having the opposite effect. Succinate oxidase showed a biphasic pattern very sensitive to KCN whereas NADH oxidase was KCN insensitive (Fig. 2b).

The affinity of the terminal oxidase located in high d.o.t. membrane towards oxygen, was tested by using an oxygen sensitive electrode in an anaerobic chamber. NADH was used as substrate for respiration.

The slopes and the intercepts of the lines were calculated from double reciprocal plots, by using groups of results from each linear correlation. Regression coefficients were highly significant ($P < 0.01$) (Fig. 3a,b). The double reciprocal plots revealed two K_m and two V_{max} values for oxygen, that indicated two pathways for oxidation of NADH with two terminal oxidases. One with high affinity to O_2, and low specific activity and the other with low affinity to O_2 and high specific activity (Fig. 3a).

Double reciprocal plots in membrane preparations amended with 1×10^{-4} M KCN revealed one straight line. The V_{max} and K_m calculated, were similar to the first oxidase with high affinity to oxygen (Fig. 3b). Similar K_m values (high affinity to O_2) with and without KCN were obtained with succinate as the oxidase substrate, but in this case V_{max} was slightly higher (not shown).

The oxygen electrode was not sensitive enough to study the kinetics of oxidases in low d.o.t. membrane particles.

Effect of Antimycin A

Antimycin A (10^{-5} mol per mg protein), completely inhibited NADH oxidase in high d.o.t. cell free extract and membranes, whereas 10^{-4} mol per mg protein did not inhibit succinate oxidase activity (Fig. 4). In low d.o.t. membranes, Antimycin A completely inhibited both NADH oxidase and succinate oxidase (not shown).

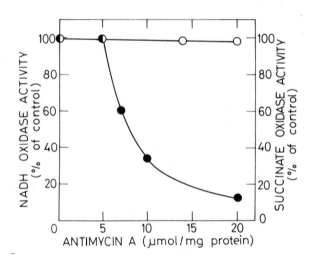

Fig. 4: Inhibition of NADH oxidase (●-●) and succinate oxidase (O-O) activities of the membrane fraction of high d.o.t. cells of A. brasilense-Cd by Antimycin A. 100% activity was 148 and 103 μLO_2/h/mg protein for NADH and succinate oxidase, respectively.

Discussion

So far, only brief attempts to characterize the electron transport system of A. brasilense have been reported (3). The results presented indicate that cell extraction by sonication yielded two membrane fractions which were readily distinguished from each other by color, structure and cytochrome content. The absence of the light fraction in low d.o.t. preparations could indicate that oxygen has a major effect on A. brasilense membranes. Furthermore in high d.o.t. membranes more proteins and more carotenoids were formed (12).

Before describing the possible electron transport pathway in A. brasilense, it is necessary to stress the effect of Antimycin A and KCN on the NADH and succinate oxidase activities in high d.o.t. membranes. Antimycin A inhibited NADH oxidase but not succinate oxidase activities. The inhibition of NADH oxidase by KCN was biphasic in the concentration rate of 5×10^{-8} and 1×10^{-3}M KCN. It was clearly observed that succinate oxidation was also inhibited by KCN, in the range of $10^{-6} - 10^{-3}$M, although this inhibition was not biphasic. In these high d.o.t. membranes, the curve using succinate as substrate can be correlated with the second inhibitory phase of NADH oxidase inhibition curve (Fig. 2a). The kinetic studies (Fig. 4a,b) also suggest that there are at least two terminal oxidases for NADH oxidation with various sensitivities to cyanide in high d.o.t. membranes of A. brasilense.

The simplest way of explaining these results is to assume that in high d.o.t. NADH and succinate oxidation proceed to oxygen through a branched pathway, with a branch point situated on the substrate side of the Antimycin A block (Cyt b 560 → Cyt c 551). According to this model, cytochrome c only donates or accepts electrons after the site of Antimycin A inhibition on the pathway which is sensitive to a low concentration of KCN, assuming that this pathway terminates in cytochrome $a-a_3$, similar to that of Rhizobium trifolii (6). The other arm of the branched chain is assumed to be blocked by a high concentration of KCN. This branch is the only pathway for succinate oxidation.

In low d.o.t. membranes it appears there are two branches for electron transport from succinate oxidation, because of the biphasic pattern of inhibition observed with KCN.

Summary of microaerobic properties

Azospirillum brasilense preferred low d.o.t. when fixing nitrogen, but also when the growth medium was supplemented with combined nitrogen (11,14). It responded to self-created and pre-formed oxygen gradient by forming aerotactic bands in capillary tubes and actively moving towards a specific zone with low d.o.t. High O_2 concentration in the capillary repulsed the bacteria. There was no band formation under anaerobic conditions, although the bacteria remained motile (2). Another means of reaching microaerobic conditions is the formation of cell aggregates (2,13). The dissolved O_2 in the band areas that forms in semi-solid agar tubes was very close to zero (2). A. brasilense utilized its energy and carbon source more efficiently for growth (with N_2 or NH_3 as nitrogen sources) under low d.o.t. (12). Also, it produced PHB in large quantities. PHB may serve as further energy and carbon sources when carbon is limiting (12).

At intermediate d.o.t. levels, A. brasilense produced red carotenoids which are apparently capable of protecting the cell against oxidative damage. (12,13). In chemostat studies, it was found that Azospirillum, readily adapted to high d.o.t. (only when combined nitrogen was present) mainly by increasing the protein content of the cells and the specific activities of succinate oxidase and superoxide dismutase (12).

Although A. brasilense contains the above mentioned protecting agents against oxygen and its radicals, it still actively seeks, by aerotaxis an environment where oxygen is limiting (microaerobic environment), this behaviour is probably less costly in energy terms than the synthesis of carotenoids and respiration enzymes. Furthermore, as shown in this work, A. brasilense cells grown in low or high d.o.t.'s possess a branch electron transport system with oxidases with high and low affinities to O_2.

The adaptability of A. brasilense to low, intermediate and high d.o.t., and the capability of rapidly seeking preferred low d.o.t. by aerotaxis, make A. brasilense a more likely candidate to be used to colonize, proliferate and survive in the rhizosphere of plants where O_2 gradients are constantly changing. It has therefore been successfuly used as inoculum for promoting growth of plants and for increasing crop yield (14).

Acknowledgements

This research was supported by a grant from the United States-Israel Binational Foundation (BSF), Jerusalem, Israel (grant No. 2476/81) and by the USA-Israel Binational Agriculture Research and Development Fund (BARD), Grant No. I-254-80.

References

1. Appleby, C.A. 1980. Biochim. Biophys. Acta 172, 88-105.
2. Barak, R., Nur, I., Okon, Y. and Henis, Y. 1982. J. Bacteriol. 52, 643-649.
3. Burris, R.H., Okon, Y. and Albrecht, S.L. 1978. Properties and Reactions of Spirillum lipoferum. In: U. Granhall (Ed.). Ecol. Bull. (Stockholm) 26, 353-363.
4. Castor, L.N. and Chance, B. 1959. J. Biol. Chem. 234, 1587-1592.
5. Cole, J.A. and Rittenberg, S.C. 1971. J. Gen. Microbiol. 69, 375-383.
6. de Hollander, A. and Stouthamer, A.H. 1980. Eur. J. Biochem. 111, 473-478.
7. Drozd, J. and Postgate, J.R. 1979. J. Gen. Microbiol. 63, 63-73.
8. Jones, C.W. and Redfearn, E.R. 1967. Biochim. Biophys. Acta 143, 340-353.
9. Lemberg, R. and Barrett, J. 1973. In: Cytochromes, pp. 217-326 Academic Press, Inc., New York.
10. Markwell, A.M., Suzanne, K., Hass, M., Bieber, L.K. and Tolbert, N.E. 1979. Anal. Biochem. 87:206-210.
11. Nur, I., Okon, Y. and Henis, Y. 1980. Can. J. Microbiol. 26, 714-718.
12. Nur, I., Okon, Y. and Henis, Y. 1982. J. Gen. Microbiol. 128, 2937-2943.
13. Nur, I., Steinitz, Y.L., Okon, Y. and Henis, Y. 1981. J. Gen. Microbiol. 122, 27-32.
14. Okon, Y. 1982. Isr. J. Bot. 31, 214-220.

15. Okon, Y., Albrecht, S.L. and Burris, R.H. 1977. Appl. Environ. Microbiol. 33, 85 -88.

16. Tissieres, A. 1956. Biochem. J. 64, 582-589.

17. White, D.C. and Sinclair, P.R. 1971. In: Advances in Microbial Physiology, Vol. 5. pp. 123-212. A.H. Rose and J.M. Wilkinson (Eds.). Academic Press, Inc., New York.

18. Yang, T.Y. and Jurtshuk, Jr. P. 1978. Biochim. Biophys. Acta 502, 543-548.

NITROGEN FIXATION (C_2H_2-REDUCTION) AND GROWTH OF PURE AND MIXED CULTURES OF AZOSPIRILLUM LIPOFERUM, KLEBSIELLA AND ENTEROBACTER sp. FROM CEREAL ROOTS IN LIQUID AND SEMISOLID MEDIA AT DIFFERENT TEMPERATURES AND OXYGEN CONCENTRATIONS

G. Jagnow

Institut für Bodenbiologie, Bundesforschungsanstalt für Landwirtschaft, Bundesallee 50, D 3300 Braunschweig, FRG

Introduction

In rhizosphere soil of temperate cereals (Pedersen et al. 1978, Jagnow 1983, 1983a) and grasses (Haahtela et al. 1981), N_2-fixing Klebsiella and Enterobacter spp. were found to be 10-100 times more numerous than Azospirillum spp.. In washed, homogenized fresh wheat roots however, populations of N_2-fixing Enterobacteriaceae and Azospirillum sp. were about equal in size with ca. $10^5.g^{-1}$ DW each in plants without N fertilizer and $10^4.g^{-1}$ each in plants given 40 or 80 $kg.ha^{-1}$ N fertilizer (Jagnow 1983a). N_2-ase activity of fresh wheat and barley roots with adhering soil when tested without preincubation varied from 1-9 nmol C_2H_4 $.g^{-1}DW.h^{-1}$ depending on fertilizer dose and time after flowering (Jagnow 1983). It is not possible at present to determine how much each of these groups may contribute to the N_2-ase activity of root samples because population sizes often are not correlated with this activity and physiological requirements are very different for each group. While Enterobacteriaceae depend on anaerobic sugar fermentation, Azospirilla depend on the microaerobic oxidation of organic acids and sugars (Tarrand et al. 1978), some may also oxidize H_2 as energy source for N_2-fixation (Watanabe et al. 1982). Their temperature responses are also likely to be different. Positive interactions resulting in increased N_2-ase activity and cell yield in the rhizosphere might therefore be possible, e.g. by utilization of fermentation products by Azospirillum sp., which was tested in model experiments with pure and mixed cultures in liquid and semisolid media with glucose and malate as energy sources under different partial pressures of oxygen. Temperature responses of C_2H_2-reduction were determined with representative pure cultures.

Materials and methods

The experimental strains were an <u>Azospirillum lipoferum</u> isolated from maize roots (Jagnow 1981), a <u>Klebsiella</u> and an <u>Enterobacter</u> sp. isolated from wheat roots and characterized previously (Jagnow 1983). For the experiments with pure and mixed cultures a liquid and semisolid malate medium (Okon et al. 1977) was supplemented with 70 ug.ml^{-1} yeast extract, in which half of the malate was replaced by glucose (glucose-malate medium). Strains were maintained on nutrient agar slants.

Temperature responses of N_2-ase activity were determined in test tubes with 10 ml semisolid malate agar for <u>Azospirillum</u> or anaerobically in semisolid glucose agar (glucose instead of malate) for Enterobacteriaceae. Tubes were inoculated with 0.1 ml of saline suspensions from 24 h slants, after this glucose agar tubes were evacuated 3 times to ca. 10 mm **Hg** and refilled with purified N_2 using rubber caps with hypodermic needles inserted. For each temperature and incubation period 3 replicate tubes were tested by injecting 1 ml C_2H_2 and by analyzing 1 ml gas samples for C_2H_4 after 2 h at the respective temperatures (Jagnow 1983). Tubes were incubated for 1,2,3,4,5,7, 10 and 14 days at 15 and 20°C in cooled and at 25, 30 and 35°C in normal incubators.

For liquid culture experiments fresh 24 h slants were suspended in 5 ml 0.3% saline and 0.5 ml inoculated into duplicate 200 ml conical flasks with 50 ml glucose-malate medium. For mixed cultures suspensions from slants of 2 organisms were mixed in equal parts and 0.5 ml was used as mixed inoculum. In some experiments flasks were incubated under reduced O_2 concentrations of 2, 0.2 and 0.02%. These flasks were closed with rubber caps to which hypodermic needles were inserted, plugged with some cotton wool in their openings before sterilization. Oxygen was reduced to 10, 1 or 0.1% of ambient pressure by evacuating 1, 2 or 3 times and refilling with purified N_2. The inocula were added through the rubber caps using disposible syringes. Flasks were shaken with 80.min^{-1} at 25°C on a linear incubation shaker or incubated as stationary cultures for 4-7 days. At daily intervals 2 flasks per treatment were removed and and acetylene reduction was determined by adding 10 ml C_2H_2 and by analyzing 1 ml samples after 2 h incubation. Extinction was measured at 540 nm and viable counts were made by spreading 0.1 ml of suitable dilutions in 0.3% saline on triplicate nutrient agar or malate agar plates with yeast extract and brome thymol blue indicator (Döbereiner et a. 1976) for Klebsiella and Azospirillum, respectively. Cells were centrifuged, washed with 1/15 m phosphate buffer of pH 7.0 and their protein contents were determined after Lowry as modified for whole cells using bovine serum albumine as a standard (Drews 1968).

Pure and mixed cultures of Azospirillum and Enterobacter in semisolid glucose-malate agar incubated under air, 2 and 0.2% O_2 for 1,2,3,4,5,7, 11, 14 and 21 days at 25°C were set up with 4 replicates and analyzed for N_2-ase activity as described. In one type of mixed culture Azospirillum grew in pure culture for 3 days before Enterobacter inoculum was added. Enterobacter pure cultures were incubated under N_2. For protein determinations culture tubes were molten in a hot water bath, precipitated with 1 ml of 20% w/v TCA and mixed with a Vortex shaker. After rinsing with ca. 40 ml of deionized hot water precipitates were centrifuged, washed in 1/15 m phosphate buffer of pH 7.0 and taken up in phosphate buffer. Protein determinations were made as described above.

Results and discussion

1. Temperature responses of nitrogenase activity

All of the three species of rhizosphere bacteria had maximal activities at 30°C, but for Azospirillum the maximum and the distribution of activity over the incubation period were distinctly different from Klebsiella and Enterobacter (table 1). Between

Table 1: N_2-ase activity (nmol $C_2H_4 \cdot tube^{-1} \cdot h^{-1}$) of cereal rhizosphere bacteria at different growth temperatures in semisolid malate or glucose agar incubated under air or nitrogen, respectively after different incubation times

	°C	1d	2d	4d	7d	10d
Azospirillum	15	0	0	0	4.4	<u>20.0</u>
Klebsiella		0.2	5.5	<u>12.1</u>	5.5	2.3
Enterobacter		1.5	6.8	<u>23.2</u>	1.8	0.3
Azospirillum	20	0	0	8.3	<u>35.4</u>	34.0
Klebsiella		10.3	<u>15.1</u>	14.5	0.5	0
Enterobacter		<u>19.2</u>	16.8	12.3	0.9	0.1
Azospirillum	25	0	13.8	<u>74.1</u>	63.1	64.1
Klebsiella		11.9	<u>41.8</u>	10.5	0.3	0
Enterobacter		<u>30.6</u>	26.4	11.3	0.4	0.2
Azospirillum	30	0	42.6	61.1	61.6	<u>77.5</u>
Klebsiella		28.4	<u>56.8</u>	6.8	0.1	0
Enterobacter		28.3	<u>32.7</u>	4.1	0.1	0
Klebsiella	35	2.2	<u>11.3</u>	6.6	6.1	1.5
Enterobacter		3.9	<u>6.5</u>	4.3	0	0

Table 2: Cumulative N_2-ase activity (nmol $C_2H_4 \cdot 14d^{-1}$) of <u>Azospirillum lipoferum</u>, <u>Klebsiella</u> sp. and <u>Enterobacter</u> sp. at different growth temperatures in semisolid malate or glucose agar under air or nitrogen, respectively

°C	15	20	25	30	35
Azospirillum	3 401	8 786	14 074	<u>14 326</u>	0
Klebsiella	1 387	1 423	2 071	<u>2 582</u>	1 001
Enterobacter	1 654	1 687	<u>2 304</u>	1 932	511

20 and 30°C its maxima were about 2 times higher than those of the Enterobacteriaceae and also remained high for longer periods:

7-14 days (14 d not shown) at 20°C, 4-14 days at 25 and 30°C, respectively. Within this temperature range, maximal activity of Enterobacteriaceae was attained during the first 2 days and was followed by a sharp drop. The reason seems to be the more efficient oxidative metabolism of <u>Azospirillum</u> and its ability to utilize both substrates and the quick conversion of glucose to acid fermentation products by <u>Klebsiella</u> and <u>Enterobacter</u> which seem to inhibit further activity. At 35°C <u>Azospirillum</u> did not grow and the activity of the Enterobacteriaceae declined strongly, while at 15°C all strains grew well and had maximal activities after 4 and 10 days, respectively. In contrast with their pathogenic counterparts the temperature range of activity is more adapted to temperate soils, the optimum being nearly attained at 25°C. There were also differences between <u>Klebsiella</u> and <u>Enterobacter</u>, the latter being more active at low, the former at high temperatures. These features can be seen more clearly in the cumulative C_2H_4-formation during 14 days (table 2).

2. Growth and nitrogenase activity in liquid cultures

Viable <u>Azospirillum</u> counts in shaken pure cultures under 2% O_2 were higher than in <u>Azospirillum-Klebsiella</u> mixed cultures during the first 2 days, but lower after the 3rd day. In mixed shaken cultures <u>Azospirillum</u> counts were higher under air than under 2% O_2, while <u>Klebsiella</u> counts were higher under lower O_2 concentration after 2 days (table 3). After 3 days <u>Klebsiella</u> almost vanished in pure culture under N_2, in mixed culture under air or with 2% O_2, but had reached its maximum in pure culture already after 1 day. If extinction is considered as a measure of cell yield, the mixed culture under 2% O_2 was more efficient by ca. 33% after 3 days than the <u>Azospirillum</u> pure culture despite the fact that <u>Klebsiella</u> counts in mixed cultures were always much lower than <u>Azospirillum</u> counts. Therefore, the yield increase of the mixed culture was probably caused by a yield increase of <u>Azospirillum</u> rather than <u>Klebsiella</u> cells.

Protein contents and spacific N_2-ase activity of pure <u>Azospirillum</u> and mixed <u>Azospirillum-Enterobacter</u> as shaken or

Table 3: Plate counts (**v. count**, $10^6 \cdot ml^{-1}$) and extinction (E_{540}) of pure and mixed shaken cultures of <u>Azospirillum lipoferum</u> and <u>Klebsiella</u> sp. in glucose-malate solution at $25°C$ (Values with different index letters differ significantly with $P = 0.05$ or less within incubation periods, as determined by the t-test)

days		Azospirillum lipoferum			Klebsiella		
		pure	mixed		pure	mixed	
		$2\% \, O_2$	$20\% \, O_2$	$2\% \, O_2$	N_2	$20\% \, O_2$	$2\% \, O_2$
1	v. count	316_a	230_b	227_b	706_c	21	1
	E_{540}	0.200	0.166	0.192	0.038		
2	v. count	363_a	275_b	223_c	28_d	22	69
	E_{540}	0.280	0.240	0.249	0.106		
3	v. count	189_a	280_b	200_a	0.3_c	1	1
	E_{540}	0.238	0.216	0.316	0.046		

stationary cultures are given in table 4. Under both O_2 concentrations, protein yield of <u>Azospirillum</u> was higher during the first 2 days in shaken pure cultures, but increasingly higher thereafter in stationary pure cultures. Specific N_2-ase activity (s.a.), on the other hand, was always much higher in stationary cultures. In stationary cultures the motile cells may be favoured by the possibility of selecting optimal positions in an O_2 gradient in the medium, especially in the later growth stages where nutrients may be limiting. While there was not much difference in yield and s.a. between 0.2 and 0.02% O_2 in the shaken cultures, in stationary cultures s.a. was higher under 0.2% O_2 during the first 3 days, but after this higher under 0.02% O_2. This is in agreement with Okon et al. (1977) which also reported highest <u>Azospirillum</u> yields if aerated with 0.01-0.02% O_2.

The stationary mixed cultures had similar or lower yields than the stationary pure <u>Azospirillum</u> cultures, but generally a much higher s.a. in the beginning, which soon decreased

Table 4: Cell protein formed (µg) and specific N_2-ase activity (s.a.: nmol C_2H_4·ug protein^{-1}·h^{-1}) in pure shaken (shake) and stationary (stat.) cultures of <u>Azospirillum lipoferum</u> and in pure and mixed stationary <u>Enterobacter</u> and <u>Enterobacter-Azospirillum</u> cultures in glucose-malate liquid medium under different O_2 concentrations during 7 days (25°C, for index letters compare table 3)

		Azospirillum lipoferum pure cultures				mixed cultures		Entero-bacter pure
		%O_2				%O_2		N_2
		0.2		0.02		0.2	0.02	
d		shake	stat.	shake	stat.	stat.	stat.	stat.
1	µg	138$_a$	56$_b$	139$_a$	68$_{ab}$	50$_b$	47$_b$	28$_c$
	s.a.	0	2 785	184	321	37 642	34 659	111 138
2	µg	210$_a$	175$_a$	209$_a$	186$_a$	261$_a$	252$_a$	60$_b$
	s.a.	77	3 284	0	1 393	10 321	15 159	2 847
3	µg	241$_a$	313$_b$	245$_a$	288$_{ab}$	230$_b$	236$_b$	57$_c$
	s.a.	98	1 292	64	972	182	105	295
4	µg	245$_a$	446$_b$	226$_a$	382$_b$	238$_a$	246$_a$	62$_c$
	s.a.	111	608	116	1 015	71	69	271
7	µg	248$_a$	541$_b$	240$_a$	664$_b$	-	-	-
	s.a.	127	362	51	716			

sharply. A still much lower yield and a much higher initial s.a. was observed in the <u>Enterobacter</u> pure cultures. The quick loss of viability of <u>Klebsiella</u> in liquid cultures and the quick drop of N_2-ase activity of <u>Enterobacter</u> and <u>Klebsiella</u> in the temperature experiment suggest the possibility of protein degradation by **autolysis**. Soluble protein resulting from this, however, was not determined since only cetrifuged, washed cells were analyzed. In this case the total N fixed may be higher in comparable mixed cultures, which seems to be indicated by the higher N_2-ase

Table 5: Cell protein formed (µg) and specific N_2-ase activity (s.a.) in semisolid glucose-malate medium under air and reduced oxygen concentration grown with pure and mixed Azospirillum (A) and Enterobacter (E) during 21 days (for index letters compare table 3)

d		A %O_2 20	A %O_2 0.2	A + E after 3d %O_2 20	A + E after 3d %O_2 0.2	mixed %O_2 20	mixed %O_2 0.2	E(N_2)
1	µg	21_a	48_b	-	-	138_c	195_{cd}	220_d
	s.a.	24_a	43_b	-	-	144_{cd}	230_c	92_d
2	µg	31_a	65_a	48_a	89_b	147_b	115_b	187_c
	s.a.	561_a	949_b	219_{cf}	716_d	86_e	139_f	37_g
3	µg	86_a	167_b	62_a	119_b	166_b	163_b	100_{ab}
	s.a.	4345_a	675_b	615_b	920_b	87_d	248_c	25_e
4	µg	215_a	272_b	278_{ab}	288_b	155_c	156_c	140_c
	s.a.	465_a	596_a	306_b	429_a	179_c	152_c	2_d
5	µg	732_a	541_{bcd}	669_{ab}	756_a	504_c	526_c	570_d
	s.a.	185_a	154_a	145_a	166_a	227_a	192_a	1_b
7	µg	876_{ab}	906_{ab}	807_a	1024_b	900_a	836_b	302_c
	s.a.	114_a	147_b	108_a	154_b	141_b	210_c	1_d
11	µg	1222_a	1567_b	958_c	1366_d	963_c	1196_a	-
	s.a.	80_a	74_a	69_a	102_b	80_a	129_c	-
14	µg	1198_{ab}	1438_{ab}	1048_{ab}	1511_{ab}	1151_a	1419_b	-
	s.a.	59_{abc}	74_c	38_a	56_b	33_a	77_c	-
21	µg	1466_a	1745_b	1304_a	1953_c	1161_a	1802_b	-
	s.a.	25_a	51_b	29_a	15_{ac}	17_c	10_c	-

activity. The maximal NH_4^+-N levels, however, were only 1.7 and 1.4 ug.ml^{-1} in stationary pure Azospirillum and in mixed cultures, respectively and the pH of culture solutions did not exceed 7.2. In the mixed cultures, it dropped to 6.1 in 2 days and gradually rose to 6.4-6.5 again. Therefore, only little protein degradation could have occurred.

3. Growth and nitrogenase activity in semisolid cultures

In contrast with liquid cultures and despite their smaller volumes, higher cell protein levels of 1 and 1.5-2 mg were obtained in semisolid glucose-malate agar cultures after 7 and 21 days, respectively. Furthermore, the highest s.a. was observed here in Azospirillum pure cultures (table 5). In these, more protein was formed under 0.2% O_2 than under air at nearly all incubation periods. Specific N_2-ase activity, however, was not always higher and after 3 days it reached its absolute maximum under air, probably being enhanced by the steeper O_2-gradient during logarithmic growth. In Azospirillum pure cultures with Enterobacter added 3 days later, not only cell yields but also s.a. were generally higher under 0.2% O_2. With some delay this was also true in the mixed cultures. Parameters for cultures under 2% O_2 were generally intermediate and therefore had been omitted.

Enterobacter cultures under N_2 and also mixed cultures under 0.2% O_2 had higher protein contents than Azospirillum cultures during the first 2 days. After this, protein content in Enterobacter cultures decreased and increased again with a second maximum of 570 µg after 5 days and a further strong decrease after 7 days. During this, s.a. was too low to support growth. These decreases again are likely to be due to autolysis. After 2 days in cultures with both organisms, however, only net cell growth was observed. While under air cell yields of pure and both of the mixed Azospirillum cultures were nearly equal, under 0.2% O_2 they often were significantly higher in Azospirillum cultures with Enterobacter added later, pointing to an increased efficiency if both organisms were cooperating under microaerobic conditions with suitable spatial separation, this is, with Azospirillum colonizing its optimal place in the O_2 gradients first with its typical pellicle and Enterobacter colonizing and fermenting glucose in the deeper layers later. Acid

Table 6: Cumulative amounts of reduced acetylene ($\mu m\ C_2H_4$) and molar ratios of reduced acetylene and cell protein-N (mol C_2H_4/N_p) in semisolid and liquid glucose-malate media inoculated with <u>Azospirillum lipoferum</u> (A), with successive inoculation: <u>Azospirillum</u> first and <u>Enterobacter</u> 3 days later (A+E), with mixed inoculum of both organisms (M) and with <u>Enterobacter</u> under N_2 (E) and kept under atmospheres with decreasing oxygen concentrations at 25°C (for index letters compare table 3).

%O_2		semisolid agar, 21 d			%O_2	liquid stationary, 7 d		
		A	A+E	M		A	M	E (N_2)
20	$\mu m\ C_2H_4$	43.0$_{ab}$	26.5$_d$	24.0$_d$	0.2	45.5	108.2	
	mol C_2H_4/N_p	6.6	4.6	4.6		18.9	102	
0.2	$\mu m\ C_2H_4$	55.3$_c$	48.9$_b$	41.3$_a$	0.02	56.8	131.6	74.2
	mol C_2H_4/N_p	7.1	5.6	5.1		19.2	120	61

formation in semisolid media under the <u>Azospirillum</u> pellicle could be shown only in mixed cultures by preparing the media with additions of brome thymol blue as pH indicator.

4. Apparent N_2-ase efficiency in pure and mixed cultures in semisolid and liquid media

From rates of C_2H_2 reduction cumulative C_2H_4 production was calculated for the pure and mixed semisolid and liquid cultures under different O_2 concentrations. From cell protein yields, molar ratios of C_2H_4 formed and protein-N fixed were calculated, assuming a N content of 6.25% for protein (table 6). The lowest molar ratios were obtained from the mixed cultures in semisolid agar: 4.6-5.6 compared with 6.6-7.1 for <u>Azospirillum</u> pure cultures, depending on the O_2 level. This demonstrates a higher N_2-ase afficiency of mixed cultures in semisolid media, approaching a theoretical optimal ratio of 4.0, which was postulated by Schrauzer (1975), because of the release and decay of the first unstable N_2-reduction product $(NH)_2$ into N_2, H_2 and N_2H_4, only the latter being reduced to 2 NH_3.

In the liquid medium, however, mixed cultures had a much lower apparent efficiency, presumably by promoting cell lysis. Considering the mucigel of the rhizoplane and the pectin layers of cell walls as ecological niches for N_2-fixing rhizosphere bacteria, cultures in semisolid agar are likely to provide a better model for the cooperation of <u>Azospirillum</u> spp. and facultative anaerobes during associative N_2-fixation in the rhizosphere.

References

Döbereiner, J., Marriel, I.E. and Nery, M. 1976: Can. J. Microbiol. 22, 1464-1473

Drews, G. 1968: Mikrobiologisches Praktikum für Naturwissenschaftler, Springer-Verlag, Berlin

Haahthela, K., Wartiovaara, T., Sundman, V. and Skujins, J. 1981: Appl. Envir. Microbiol. 41, 203-206

Jagnow, G. 1981: Experientia Suppl. Vol. 42, 100-107

Jagnow, G. 1983: Z. Pflanzenernaehr. Bodenk. 146, 217-227

Jagnow, G. 1983a: Rev. d'Écol. Biol. Sol, spec. Vol.: Proc. 8th Internat. Coll. of Soil Zoology, Louvain-la Neuve 1982

Okon Y., Albrecht, S.L. and Burriss, R.H. 1977: Appl. Envir. Microbiol. 33, 85-88

Pedersen, W.L., Chakrabarty, K., Klucas, R.V. and Vidaver, A.K. 1978: Appl. Envir. Microbiol. 35, 129-135

Schrauzer, G.N. 1975: Angew. Chemie 87, 579-587

Tarrand, J.F., Krieg, N.R. and Döbereiner, J. 1978: Can. J. Microbiol. 24, 967-980

Watanabe, I., Barraquio, W.L. and Daroy, M.L. 1982: Can. J. Microbiol. 28, 1051-1054

ECOLOGICAL FACTORS AFFECTING SURVIVAL AND ACTIVITY OF AZOSPIRILLUM IN THE RHIZOSPHERE

S. L. ALBRECHT, M. H. GASKINS, J. R. MILAM, S. C. SCHANK and R. L. SMITH

USDA-ARS, Departments of Agronomy and Microbiology and Cell Science, University of Florida, Gainesville, FL 32611, USA

Introduction

Nitrogen-fixing bacteria which grow in association with crop plant roots have the potential to reduce nitrogen fertilizer requirements in many agricultural areas. Inoculation of these bacteria into the rhizosphere to augment or initiate nitrogen-fixing systems has been studied extensively (2, 8, 9). Many of these also produce plant growth hormones which in some situations increase growth rates of the plants. The growth stimulating effect of the hormones produced may be of sufficient importance in some circumstances to justify bacterial inoculation.

Yield increases after inoculation with A. brasilense have been found in many places throughout the world. Inoculation has produced substantial yield increases in many cases, but responses of treated plants are highly unpredictable for reasons that remain obscure. This unpredictability has discouraged commercial exploitation. Some recent reports of consistent responses to inoculation with A. brasilense in Israel (6) have renewed interest in this research. Most investigators are interested in the capacity of the bacteria to fix nitrogen, but some are studying plant responses to the growth-promoting hormones produced by the bacteria.

The assumption that bacteria added to the rhizosphere will survive, grow and remain metabolically active is not justified by most experimental results. It is essential to maintain large numbers of nitrogen-fixing bacteria in the rhizosphere to provide nitrogen fixation rates significant for agricultural needs. This may not be possible until further information about their requirements in the soil is obtained.

After inoculation, survival of the bacteria is rarely monitored, and most investigators who attempt to reisolate the inoculated organism use only enrichment culture methods. This procedure reveals little or nothing about the abundance of the organism or the metabolic activity in the rhizosphere. Our data indicate that when Azospirillum or similar species are inoculated into the soil, their numbers decline rapidly. Their behavior in the soil is similar to that of fecal coliform bacteria, some of which are morphologically and metabolically similar to the nitrogen-fixing species. It has been shown that populations of these organisms decline rapidly after introduction into natural environments (4, 5).

This report describes a series of experiments in which populations of nitrogen-fixing bacteria were monitored following soil inoculation, to determine factors that affect their numbers and to evaluate their effects on several crop plants.

Materials and Methods

Experiment I

Azospirillum brasilense strain 13t was inoculated into flasks containing mixtures of sterile (gamma-irradiated) sand and peat (neutralized with $CaCO_3$). Moisture was adjusted to 40 percent (w/w) and the flasks were incubated at 25 C. Samples of 10 g were removed at selected intervals and dilution plate counts were made on nutrient agar.

Experiment II

Sorghum (Sorghum bicolor) plants were grown in a non-sterile Kendrick fine sand (loamy, siliceous, hyperthermic Arenic Paleudult) in a greenhouse during the late spring. Natural daylength was used and the maximum irradiance was 1500 µE m^{-2} sec^{-1}. Temperature ranged from 25 to 35 C, and mean relative humidity was 60 percent. The soil pH was adjusted to 6.3 with lime. The surface of the soil was covered with washed dark gravel to suppress growth of cyanobacteria. Plants were inoculated with a liquid suspension of A. brasilense strain 13tSR2 (a double-marked antibiotic mutant resistant to streptomycin and rifampicin), supplied by D. E. Duggan. The bacteria were grown and prepared for inoculation as described earlier (1).

The water capacity of the soil was determined by drying a saturated sample to constant weight. Three moisture treatments were then selected to correspond to 90, 70 and 50 percent of saturated conditions. Pots were weighed daily, and each was adjusted to its assigned weight by addition of distilled water. As the plants became larger and the days warmer, soil moisture adjustments were made twice daily.

Nitrogenase activity was estimated by the core method of Tjepkema and Burris (11). Cores were incubated at ambient temperatures for 24 hours in the dark. Then the dry weights of shoots and roots were determined after drying to a constant weight at 60 C. Numbers of soil bacteria were determined by the agar-plate method. Aerobic heterotrophs were counted on trypticase soy agar incubated at 28 C for 48 hours. The indigenous population of Azospirillum-type organisms was counted on malate nitrogen-free agar (MNF) and the numbers of the introduced 13tSR2 strain were counted on Wood's agar (A. G. Wood and M. E. Tyler, personal communication).

Experiment III

A. brasilense strain CdSR (a mutant resistant to streptomycin and rifampicin) was inoculated into either the rhizosphere of Zea mays growing in washed quartz sand or soil (Kendrick fine sand), or into soil or sand without plants. Plants were grown in a greenhouse under natural light and daylength. At selected intervals, 10 g samples were taken from the pots and the A. brasilense populations were estimated by the most probable number (MPN) method in MNF broth, with the appropriate antibiotics. Total bacterial numbers were estimated by MPN using thioglycolate broth medium.

Experiment IV

A field experiment was planted with two species, pearl millet (Pennisetum americanum, variety 'Tifleaf') and an interspecific hybrid of pearl millet (inbred variety 23 DA) and napiergrass (P. purpureum Schumach, strain N75).

The soil, a well-drained Arredondo fine sand (siliceous, hyperthermic Grossarenic Paleudult), had a pH of 6.5. Four nitrogen treatments (main plots) were used: 0, 30, 60 and 120 kg N ha^{-1}. The plots consisted of single 0.9 m wide rows, 4.6 m long. A split-plot design, replicated 10 times, was used. Fertilizer N treatments were randomized within each block and separated by border rows. Plots were irrigated weekly and weed and insect control treatments were applied as needed. The plant species were arranged in pairs, and located in nitrogen fertilizer plots so that three pairs were inoculated with either Cd, CdSR or autoclaved inoculum as required by the design. Before inoculation, the bacteria were mixed with a 10 percent peat carrier previously neutralized with $CaCO_3$. Approximately 1.3×10^7 cells were applied per cm of row. Counts of the Azospirillum strains were made using the MPN method. Succinate N-free medium was used, incorporating the appropriate antibiotics.

Results

Figure 1 shows the survival of A. brasilense strain 13t in mixtures of peat and sand. After the initial decline most populations, with the exception of those on the 0.5% peat-content media, remained stable for about 60 days. With few exceptions, throughout the course of the experiment those systems with the greatest peat content had the greatest populations of A. brasilense. Populations remained very high during the experiment, comparable to other systems with A. brasilense as the sole inhabitant. In experiments where A. brasilense was not the only bacterium or the predominant species present, they were readily lost from the system.

The nitrogenase activity associated with inoculated A. brasilense strain 13tSR2 is presented in Figure 2. Activity of the cores was very low until the seventh week of the experiment, when it increased to a maximum of 38 nmoles core^{-1} hour^{-1} in those pots maintained at 90 percent of field capacity. Nitrogenase activity was positively correlated with soil moisture throughout the experiment. Bacterial populations in the rhizosphere are shown in Figure 3. Populations of the marked mutant strain decreased three orders of magnitude in about seven days. By the tenth week of the experiment, the populations of the mutant seemed to be increasing in the highest soil moisture regime and decreasing in the driest pots. At the intermediate moisture level, populations did not fluctuate from levels established by the second week.

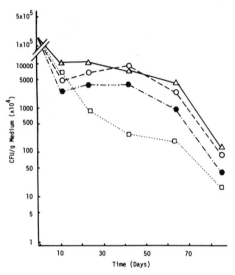

Figure 1. Survival of <u>A</u>. <u>brasilense</u> strain 13t on peat-sand mixtures. Moisture was 40 percent (w/w), pH was adjusted to 7 and storage temperature was 25 C. Bacterial populations determined by dilution plate count technique. Peat content adjusted to 0.5% (□), 1% (◐), 2% (○) and 3% (△).

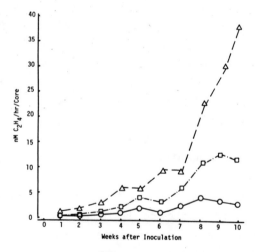

Figure 2. Nitrogenase activity (ARA) of <u>S</u>. <u>bicolor</u> roots inoculated with <u>A</u>. <u>brasilense</u> strain 13tSR2. Soil moisture maintained at 50% (●), 70% (□) and 90% (△) of field capacity.

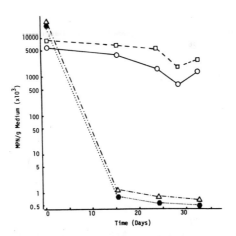

Figure 4. Survival of A. brasilense strain CdSR inoculated into soil and sand. Bacterial numbers were determined in soil (CdSR = △ ; Indigenous bacteria = □) and sand (CdSR = ● ; Indigenous bacteria = ○).

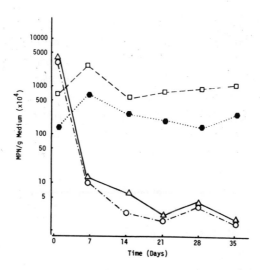

Figure 5. Survival of A. brasilense strain CdSR inoculated into the rhizosphere of Z. mays. Plants were grown in soil (CdSR = △ ; Indigenous bacteria = □) and sand (CdSR = ○ ; Indigenous bacteria = ●).

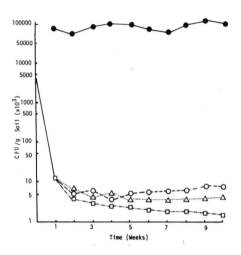

Figure 3. Survival of <u>A</u>. <u>brasilense</u> strain 13tSR2 in the rhizosphere of <u>S</u>. <u>bicolor</u>. Plants grown in a glasshouse. Soil moisture maintained at 50% (□), 70% (△) and 90% (○) of field capacity. Indigenous aerobic heterotrophic bacteria shown for 70% (●) field capacity.

Figures 4 and 5 show population dynamics of strain CdSR inoculated into soil and the rhizosphere, respectively. Total bacteria were always less in the sand pots. Few of the thioglycolate MPN tubes showed any anaerobic growth, suggesting that the majority of the bacteria in the soil were aerobic. The numbers of strain CdSR were reduced to a lesser extent in the rhizosphere, indicating that a growing plant is beneficial for the introduced bacteria.

Second harvest dry matter yields of the 'Tifleaf' pearl millet and the hybrid are shown in Figure 6. The hybrid inoculated with strain Cd responded to inoculation with significantly increased dry matter yields at the 30 and 60 kg N ha^{-1} fertilizer levels (p=0.006 and p=0.02 respectively). At the 30 kg N ha^{-1} rate the Cd inoculated plots yielded 24 percent more dry forage than the control, and at the 60 kg N ha^{-1} rate, the increase was 18 percent. The CdSR inoculated plots fertilized with 30 kg N ha^{-1} yielded 16 percent more dry matter than the controls (significant at p=0.11). Inoculation with strain Cd also increased the total amount of nitrogen harvested from those plots giving dry matter responses to inoculation. The first harvest and the pearl millet of the second harvest did not respond to inoculation.

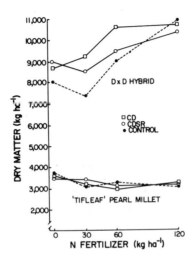

Figure 6. Dry matter yields (in kg ha^{-1}) of pearl millet variety 'Tifleaf' and the Pennisetum DxD hybrids at four nitrogen fertilizer levels.

Figure 7 shows the decline of strains Cd and CdSR over the course of the experiment. Each point on the plot is a mean of four counts from pooled soil samples. Both strains had fallen to about 5×10^2 bacteria per g soil by the sixth week after inoculation. Ten times as many indigeneous N_2-fixing bacteria were present. No Cd or CdSR strains were found in the control plots throughout the season. Sampling to determine inoculum movement through the soil six weeks after inoculation showed that little movement had occurred. By the seventh week after inoculation, strains Cd and CdSR were barely detectable. By spring 1983 they could not be recovered.

Discussion

Little is known about the survival or growth of nitrogen-fixing organism in the rhizosphere following inoculation. The ecology of associative nitrogen-fixing bacteria has been given little attention. Undoubtedly, in native soils these bacteria can survive in their niches for long periods of time. In an analogous situation, Lowendorf (7) reports that rhizobia have been shown to survive for as long as 125 years. Our studies in Florida show clearly that associative nitrogen-fixing bacteria are ubiquitous, but Azospirillum are present only in very low numbers.

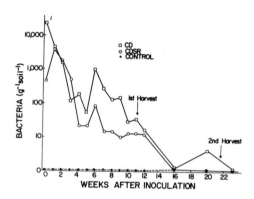

Figure 7. Bacterial populations in the field rhizosphere. Most probable number counts of strain Cd, CdSR and the control. A weekly sampling schedule was used for the first 12 weeks after inoculation, less frequently thereafter. Each point represents four soil samples.

Our results show that lack of adequate soil moisture can have deleterious effects on some strains of associative bacteria. This factor is important because it may pose a substantial barrier to the survival of inoculated organisms, especially in regions of low soil moisture due to scant rainfall and few facilities for irrigation.

Survival of the inoculant organisms was greater when growing plant roots were present in the rooting media. This rhizosphere effect suggests that bacteria obtain a substantial benefit, probably nutritional, from the roots. Extended survival was found in rooting media containing added organic matter. It is interesting to speculate that some strains of associative nitrogen-fixing bacteria may have the ability to use native soil organic matter as a source of nutrients. It is a distinct possibility that a microbial association is formed, in which microorganisms capable of metabolizing complex organic material produce compounds easily metabolized by the nitrogen-fixing populations. However, it is also possible that the function of the increased organic matter is to provide a suitable habitat by increasing moisture content, supplying mineral elements required for inorganic nutrition, or providing buffering capacity or some similar improvement of the environment. The results suggest that survival of inoculated nitrogen-fixing bacteria will be facilitated when the organisms are placed in the rhizosphere of growing plants and when the soil organic matter is high.

Yield increases of 24 percent dry matter and 39 percent total nitrogen (at the 30 kg N ha^{-1} rate) due to inoculation with Cd were obtained, and if consistent, would be very important in commercial agriculture. The inconsistency of responses in the field experiment, where only one of the two crops responded, is similar to our previous results. Inoculation responses appeared to be affected by the rate of nitrogen fertilizer application, as found previously (9).

The rapid decline of the Azospirillum populations suggests that these organisms are not competitive when introduced into the rhizosphere. Measured acetylene reduction activity was low and variable and was not influenced by inoculation, indicating nitrogen fixation rates not above 3 kg ha^{-1} per season. This suggests that nitrogen fixation by the inoculated organism is not the main mechanism in the stimulation of crop yield by A. brasilense. Other mechanisms for the stimulation of plant yield by nitrogen-fixing soil bacteria have been advanced. Barea and Brown (3) showed that Azotobacter can produce gibberellic acid, indole-3-acetic acid and cytokinin, which can modify plant growth. Tien et al. (10) demonstrated that Azospirillum can also produce growth promoting hormones, and proposed that those substances were responsible for inoculation responses.

Those results indicate that the normal course of events after introduction of nitrogen-fixing bacteria into the soil or rhizosphere is for their numbers to decrease rapidly. Survival, although improved by high levels of soil moisture and organic matter and by the presence of growing plant roots, is nevertheless generally poor. We believe frequent failure of the inoculant strain to become established in the rhizosphere is responsible for inconsistent plant responses. Until the cause of this inconsistency is identified and eliminated, the prospect for commercial use of Azospirillum inoculation to improve crop performance seems limited.

Acknowledgement

This research was supported by USAID contract AID/ta-C-1376, the USDA-ARS and the Florida Agricultural Experiment Stations. The views and interpretations in this publication are those of the authors and should not be attributed to the Agency for International Development of any individual acting in its behalf. We thank Leslie Villarreal, Douglas Manning, Kenneth Cundiff, Anthony Bouton, Paul Merritt and Kelly Kirkendall for technical assistance.Our results show that lack of adequate soil moisture can have deleterious effects on some strains of associative bacteria. This factor is important because it may pose a substantial barrier to the survival of inoculated organisms, especially in regions of low soil moisture due to scant rainfall and few facilities for irrigation.

References

1. Albrecht, S. L. and Y. Okon. 1981. Methods in Enzymology. 69C:740-749.
2. Barber, L. E., J. D. Tjepkema, S. A. Russell and H. J. Evans. 1976. Appl. Environ. Microbiol. 32:103-108.
3. Barea, J. M. and M. E. Brown. 1976. J. Appl. Bacteriol. 37:583-593.
4. Bell, R. G. and J. D. Bole. 1976. J. Environ. Qual. 5:417-418.
5. Elliott, L. F. and J. R. Ellis. 1977. J. Environ. Qual. 6:254-251
6. Kapulnik, Y., J. Kigel, Y. Okon, I. Nur and Y. Henis. 1981. Plant and Soil 61:65-70.
7. Lowendorf, H. S. 1980. Adv. Microbial Ecology. 4:87-117.
8. O'Hara, G. W., M. R. Davey and J. A. Lucas. 1981. Can. J. Microbiol. 27:871-877.
9. Smith, R. L., J. R. Bouton, S. C. Schank, K. H. Quesenberry, M. E. Tyler, J. R. Milam, M. H. Gaskins and R. C. Littell. 1976. Science 193:1003-1005.
10. Tien, T. M., M. H. Gaskins and D. H. Hubbell. 1979. Appl. Environ. Microbiol. 37:1016-1024.
11. Tjepkema, J. D. and R. H. Burris. 1976. Plant and Soil 45:81-94.

FORAGE GRASSES INOCULATION WITH GENTAMICINE AND SULFAGUANIDINE RESISTANT MUTANTS OF AZOSPIRILLUM BRASILENSE.

A. Marocco (*), M. Bazzicalupo (**) and M. Perenzin (*)

(*) Istituto Sperimentale per le Colture Foraggere, viale Piacenza 25, 20075 Lodi (Italy).

(**) Istituto di Anatomia Comparata, Biologia Generale e Genetica, via Romana 17, 50125 Firenze (Italy).

Introduction

Azospirillum spp. have been found in soil and roots from all over the world (1). The Gramineae such as maize, wheat, sorghum, sugar-cane and forage grasses are most frequently cited as host (2) and seems now easy to demonstrate nitrogenase activity in a large variety of grasses (3).

Increased yield and nitrogen contents resulting from inoculation of plants with Azospirillum spp. cannot automatically be attributed to N_2 fixation by the inoculant (4). Azospirillum spp. are well known to influence plant growth through the production of hormones (5) or by affecting the rhizosphere populations (6). The effect of inoculation is known to depend, also, to choise of plant and bacteria genotypes (3, 7).

A large number of data have accumulated especially under tropical condition and in wet soils while there are less information for temperate regions characterized by high yielding forage grasses such as Dactylis sp. and Festuca sp..

One of the problems with inoculation experiments in greenhouse or in the field is the difficulty of recognize the applied bacteria. This problem can be minimized using mutants of Azospirillum spp. wich are resistant to antibiotics or with a different pigmentation. In the present work we describe two mutants of Azospirillum brasilense and their utilization in studing the agronomic effects of the inoculum with two forage grasses Dactylis glomerata L. and Festuca arundinacea Schreb.

Azospirillum brasilense strains

The strains used were: SpF 267 resistant to gentamicine and SpF 57 resistant to sulfaguanidine; both strains were spontaneous mutants of strain Sp 6 (8). Main traits of mutant SpF 267 are summarized in Table 1. The results showed that the mutation do not change the abilities to fix nitrogen and survive in the soil. Moreover, in order to test a second strain easy to isolate from the soil, mutant SpF 57 was used. SpF 57 is resistant to sulfaguanidine and also to toxic analogues of tryptophan (5-metil-tryptophan and 5-fluoro-tryptophan), and showed a marked red pigmentation. Cross-feeding experiments demonstrated that this mutant was able to excrete tryptophan.

Inoculation and bacterial counts

Azospirillum brasilense culture were grown in plates of MSP containing 20 mM NH_4Cl and incubated at 35°C. 10^8 total cells diluted in 2 liters water were used for each container (described below). The inoculation was repeted 2 times at the end of tillering and after the first cut.

Survival of bacteria in the rizosphere was followed collecting 2 plants for each treatment; sampling was repeated 2 times after the first and the second inoculation. The number of bacteria was counted both before and after sterilization of the roots with 1.5% Na hypochlorite for 30". Appropriate dilution of roots samples were plated on MSP plus 5 µg/ml gentamicine (SpF 267) or 50 µg/ml sulfaguanidine (SpF 57).

The results are shown in Table 2. SpF 267 population in non-sterilized roots was stable during the first 2 weeks, decrease at the 9th week after the first inoculum to 4×10^4 bacteria/g roots and did not decrease until the the fourth harvest time (9th week after the second inoculum). Since then the Azospirillum population was completely lost. The same population in sterilized roots had magnitude of 2×10^3 and 2×10^2 in Festuca and Dactylis respectively after 2 weeks from the first inoculum. Starting from 9th week after the same inoculum, SpF 267 in sterilized roots increased slowly.
Strain SpF 57 was not found in sterilized roots while in non-sterilized roots its population decrease more rapidly than SpF 267.

The use of resistant mutation for distinguishing between inoculated strains and other bacteria has been important. Mutations permits, in fact, to resolve the difficulty of isolation of Azospirillum in experiments in wich it is important to count inoculated bacteria in non-sterilized soil.

Effects of Azospirillum on yield and total nitrogen

Experiment was sown in greenhouse on 15.12.1982 in concrete container of 150 cm lenght, 25 cm large and 40 cm of depth. The soil was sandy with 76% sand, 21% lime, 3% clay, 3.4% of organic matter, 2.24‰ of total nitrogen and pH 6.66. Each container was divided in two plots, one of Dactylis glomerata cv. Dora and one of Festuca arundinacea cv. Manade. Each of them containing 27 plants in single row. Before sowing, each container receive the equivalent of 150 units P_2O_5/ha and 250 units K_2O/ha. Three rates 0, 60 and 120 units/ha of fertilizer nitrogen was applied as NH_4NO_3 and distribued in 2 times: half at the end of the tillering and half after the first harvest. The treatments (3 N level, 2 forage species, 2 bacterial strains and control) were carried out in randomized block design with 8 replicates. Plants of

Dactylis and Festuca were harvested four times every 30 days starting from 20.4.1983. Dry matter yield was determined after 24 hours at 105°C. Nitrogen content was determined with a Technicon nitrogen AutoAnalyzer (9).

Results on means of dry matter and total nitrogen yield are reported in Tables 3 and 4. With Festuca arundinacea cv. Manade inoculation increased significantly the dry matter yield by 11% both with mutant SpF 267 and SpF 57. Inoculation of Dactylis glomerata cv. Dora by either strains SpF 267 and SpF 57 caused an increase of 2.8% and 8.7% respectively, in dry matter yield but differences were significant only with SpF 57. A significantly higher (13%) total nitrogen yield was obtained in Festuca inoculated with the two mutants; significant increase (7.6%) was observed in Dactylis only with strain SpF 57.

The Festuca plants showed high sensitivity to inoculation treatments in this experiment with both strains. Dactylis glomerata responded positively to SpF 57 inoculation both on dry matter and total nitrogen. No differences was found for Dactylis - SpF 267 association.

Similar results were observed in extensive inoculation experiments with Azospirillum spp. carried out in greenhouse on other species of forage grasses (1). It has been suggested that the effectiveness of association could depend both on plant and bacteria genotypes (7). The results obtained with Dactylis could indicate that plant - bacteria specificity exist. In fact, a positive interaction was found only with strain SpF 57.

Inoculation increased plant dry weight mainly at low level of nitrogen fertilization and was statistically significant at 60 units N/ha (20.5%) in Dactylis with SpF 57 (Fig. 1). Significant interaction between strain SpF 267 and fertilizer nitrogen concentration was obtained in Festuca for nitrogen yield that was significant higher with 60 units N/ha (31%; Fig.2).

The results obtained are in agreement with the observation that the main effects of inoculum is found at intermediate level of nitrogen fertilization (10, 11).

The main differences between inoculated and non-inoculated plots were observed on yield at first cut. In particular, a significant interaction between strains and cuts was obtained in Festuca dry matter with SpF 57 (Fig. 3): the difference between SpF 57 and control was 20.3%. The absence of positive effects in the following cuts, could be related to the survival of bacteria in soil.

References

1. Gibson, A.H., Jordan, D.C. 1983. In: Encyclopedia of Plant Physiology 12C, 348-382.
2. Okon, Y. 1982. Isr. J. Bot. 31, 214-220.
3. Neyra, C.A., Dobereiner, J. 1977. Adv. Agron. 29, 1-38.
4. Patriquin, D.G. 1982. In: Advances in Agricultural Microbiology. Ed. Subba Rao N.S., 139-190.
5. Reynders, L., Vlassak, K. 1979. Soil Biol. Biochem. 11, 547-548.
6. Kloepper, J.W., Schroth, M.M., Miller, T.D. 1980. Phytopathology 70, 1078-1082.
7. Rennie, R.J., Larson, R.I. 1979. Can. J. Microbiology 57, 2771-2775.
8. Bani, D., Barberio, C., Bazzicalupo, M., Favilli, F., Gallori, E., Polsinelli, M., 1980. J. Gen. Microbiology 119, 239-244.
9. Ferrari, A., 1960. Am. New York Acad. Sci. 27, 792-800.
10. Smith, R.L., Bouton, J.H., Schank, S.C., Quesenberry, R.H., Tyler, M.E., Gaskins, M.H. 1976. Science 193, 1003-1005.
11. Kapulnik, Y., Sarig, S., Nur, I., Okon, Y., Kigel, J., Henis, Y. 1981. Exp. Agric. 17, 171-178.

Table 1. Characteristics of strain SpF 267 resistant to gentamicine (* nitrogenase was assayed as described (9); ** number of bacteria/g of soil are reported).

Frequency of resistant mutants on 5 µg/ml of gentamicine			2×10^{-9}
Minimal inibitory concentration of gentamicine (µg/ml)		SpF 267	30
		Sp 6 (parental)	1.25
Doubling time in MSP (9) plus 20 mM NH_4Cl (minutes)		SpF 267	105
		Sp 6	120
Nitrogenase activity * (µmoles C_2H_4/h x mg protein)		SpF 267	10.9
		Sp 6	13.6
Survival of bacteria in sterilized soil **		Sp 6	SpF 267
	0	1.6×10^7	1.6×10^7
	5	2×10^4	6×10^4
	23	4×10^4	8.6×10^4

Table 2. Azospirillum populations in the rhizosphere (number of bacteria/g fresh roots; NS = non-sterilized roots; S = sterilized roots).

Associations	Weeks after 1st inoculation				Weeks after 2nd inoculation			
	4		9		2		9	
	NS	S	NS	S	NS	S	NS	S
Festuca + SpF 267	2×10^6	2×10^3	4×10^4	$<10^2$	3×10^4	–	7×10^4	1.3×10^5
Dactylis + SpF 267	2×10^5	2×10^2	6×10^4	2×10^4	4×10^4	–	3×10^4	1.6×10^4
Festuca + SpF 57	10^6	$<10^2$	4×10^4	$<10^2$	5×10^3	–	–	–
Dactylis + SpF 57	$<10^2$	$<10^2$	10^4	$<10^2$	5×10^3	–	–	–

Table 3. Effects of inoculation on dry matter yield (means of 4 cuts; **P = 0.01; * P = 0.05).

Strains	Dry Matter Yield (grams per plot)	
	Festuca	Dactylis
SpF 267	29.5**	22.4 ns
SpF 57	29.5**	23.7*
Control	26.6	21.8

Table 4. Effects of inoculation on total nitrogen yield (means of 3 cuts; **P = 0.01; *P = 0.05).

Strains	Total Nitrogen Yield (grams per plot)	
	Festuca	Dactylis
SpF 267	3.13**	2.79 ns
SpF 57	3.12**	2.97*
Control	2.76	2.76

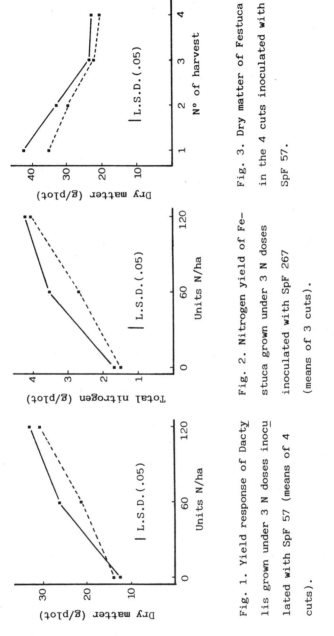

Fig. 1. Yield response of Dactylis grown under 3 N doses inoculated with SpF 57 (means of 4 cuts).

Fig. 2. Nitrogen yield of Festuca grown under 3 N doses inoculated with SpF 267 (means of 3 cuts).

Fig. 3. Dry matter of Festuca in the 4 cuts inoculated with SpF 57.

———— INOCULATED   - - - - CONTROL

ASSOCIATION BETWEEN WHEAT AND AZOSPIRILLUM LIPOFERUM UNDER GREENHOUSE CONDITIONS: INCREASE OF YIELD AND THOUSAND CORN WEIGHT

Th. Mertens and D. Heß
Lehrstuhl für Botanische Entwicklungsphysiologie
Universität Hohenheim, Emil Wolffstr.25,
7000 Stuttgart 70, FRG

Introduction

Triticum induces nitrogenase activity in Azospirillum lipoferum (Heß,Kiefer 1981). Replacing the agar medium by soil, it could be demonstrated, that Triticum induces Azospirillum on soil which was supplemented with liquid culture medium under sterile conditions. Therefor we started experiments on soil under greenhouse conditions. Growth and development of plants and especially grain yield, thousand corn weight and total compound of nitrogen of the grain were studied.

Materials and Methods

Bacteria: Azospirillum lipoferum sp108 st (Dr. J. Döbereiner, Instituto de Pescuisa Agropecuaria Centro Sul EBRAPA, Rio de Janeiro, Brasil). Plants: Triticum aestivum var. Arkas (Landessaatzuchtanstalt, Universität Hohenheim, Germany). Cultivation of bacteria: stem cultures of Azospirillum lipoferum sp108st were kept Nfb agar medium (pH 6,8, 15g agar per l; Döbereiner and Day 1975) and subcultured weekly. The strain was examined routinily for contamination by plating on the media mentioned. The inokulum was grown on semisolid media (1,75 agar per l) at 30°C in a 25 l fermentor. Cultivation of plants: Triticum was sowed in Kick-Brauckmann pots containing 8 l soil each. After germination each pot contained 20 wheat plntlets. The plants were grown up under greenhouse conditions. Associations: Four and five weeks after sowing the pots were inoculated with 250 ml medium each, containing 10^8 bacteria per ml. The controls were treated with autoclaved bacteria suspension. Grain yield and thousand corn weight were determined. The total compound of nitrogen of the grain was determined by the method of Kjeldahl.

Results and Discussion

During the growing and riping period we could not find any differences between the plants of the experimental assay and the controls This may be due to the fact, that the soil we used contains 0,28 % total nitrogen. The determination of the grain yield,however showed that the plants of the assay produced a yield which was 32,11 % higher than the controls which were treated with autoclaved bacteria suspension.

Fig.1: Increase of grain yield in assotiations with azospirilla.

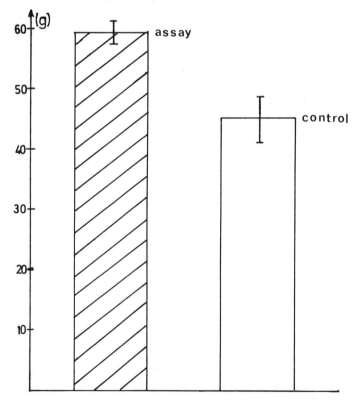

The thousand corn weight of the assay was 32 g and that of the controls only 26 g. This is an increase of 23,12 %.

Fig.2: Increase of thousand corn weight in associations with Azospirilla.

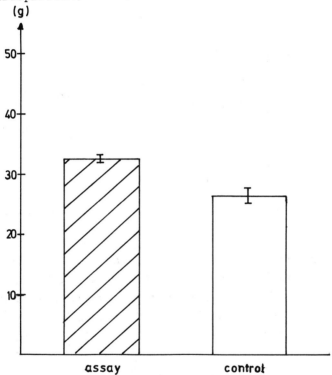

Furthermore we determined the total nitrogen compound of grain by the method of Kjeldahl. The results showed that the total compound of nitrogen of grain in the assay was 15 % significant much higher than in the controls (Fig.3). In any case an increase of grain yield was possible by inoculation with Azospirilla. This effect cannot be due to the growth factors in the medium because they were used in the controls too. It could be however that Azospirilla excrete some hormonal growth substances which influences the growth of the plants in a positive way.

Further investigations have to show wether bacterial growth

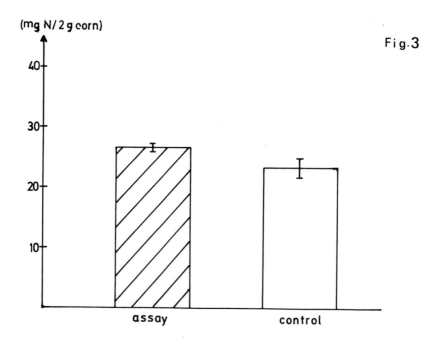
Fig. 3

substances or bacterial nitrogen fixation stimulated the increase of grain yield, thousand corn weight and of the total nitrogen compound of grain. Investigations with a different kind of soil and the influence of different amounts of nitrogen fertilizer on the association are not yet finished.

References

Heß, D. and Kiefer, S. 1981, Z. Pflanzenphys. 101, 15-24.

Döbereiner, J. and Day, J. N. 1975, in: Nitrogen Fixation by Free-living Microorganisms, pp. 39-56, ed. by W. Steward, Cambridge Univ. Press, Cambridge.

BENEFITS OF AZOSPIRILLUM INOCULATION ON WHEAT: EFFECTS ON ROOT DEVELOPMENT, MINERAL UPTAKE, NITROGEN FIXATION AND CROP YIELD.

Y. KAPULNIK and Y. OKON
Department of Plant Pathology and Microbiology
Faculty of Agriculture, The Hebrew University of Jerusalem
Rehovot 76100, Israel.

Introduction

Azospirillum has been used experimentally for inoculation of cereal and forage grass crops in both intensive and extensive agriculture. Positive effects on plant growth and increases in crop (foliage and grain) yield have been obtained (2,6,8,12,16,17). The benefit of Azospirillum to plants was mainly derived from increases in dry weight and nitrogen content of shoots, and in the average number of ears per plant in sorghum and corn (6,8,16,17).

The nitrogen fixation activity as measured by the acetylene reduction method, with detached roots and in cores with roots with immediate activity, has not been correlated with the yield obtained (1). It is widely accepted now that acetylene reduction activity by preincubated detached roots, overestimated nitrogen fixation activities.

Furthermore, no interaction was found between inoculation with Azospirillum and the level of combined nitrogen initially applied to the soil. Azospirillum enhanced the yield of several levels of N-fertilizer (6,8,16). It has been suggested that other factors such as the production of plant growth promoting substances (19) and the increase in the rate of mineral uptake by the plants (10) also contribute to the enhancement of the yield.

Effect of Azospirillum inoculation on grain yield of wheat

In temperate regions or in warmer regions with cold winters such as Israel, there are several limiting environmental factors (soil temperatures below 15°C) which can influence the capability of colonization and the activity of Azospirillum brasilense (optimum growth and nitrogen fixation activities 33-35°C).

In spite of those difficulties inoculation of wheat with Azospirillum increased plant dry matter and grain yield (Table 1).

Table 1: Wheat responses to field inoculation with Azospirillum

Reference	Location	Year	Parameter	% difference in comparison to control
Kapulnik et al. (1983) (8)	Israel	1978	grain yield	+ 11%
		1979	grain yield	+ 8%
		1979	grain yield	+ 7%
		1979	shoots dry weight	+ 19.6%
Y. Kapulnik (unpublished)	Israel	1980	grain yield	N.S.
		1981	grain yield	+ 6%
		1981	grain yield	N.S.
		1981	grain yield	+ 17.6%
		1982	grain yield	+ 10%
Reynders and Vlassak (1982) (16)	Belgium	1979	grain yield	+ 9-15%
Favilli et al. (personal communication)	Italy	1980	grain yield	+ 20%
Hegazi et al. (1980) (5)	Egypt	1980	grain yield	+ 270%
			shoots dry weight	+ 176%
Boddey and Döbereiner (1982) (2)	Brazil	1980	shoots dry weight	N.S.

N.S. No significant difference. All others showed statistically significant differences.

The better yields obtained were mainly derived from increases in tillering and in the number of fertile tillers per unit area (9,20). Moreover, mineral accumulation and content (N, P and K) in the inoculated plants was found to be higher as compared to non-inoculated controls. These benefits correlated well with the size of the plant root system (9), but not with the nitrogen fixing activity as measured in cores containing the roots (Y. Kapulnik, unpublished).

Contribution of biological nitrogen fixation

There is still controversy whether the nitrogen fixed by Azospirillum-grass associations, contributes to the nitrogen yield of the plants (1,2, 11, 12, 20).

Acetylene reduction activity was measured in detached roots, collected from several natural ecosystems in Israel, and in cores with roots from different cultivated and wild wheats growing in a net house. In all systems tested at 20°C (about 850 assays) only low acetylene reduction activities (50-100 nmole C_2H_4 produced/h/g dry weight of roots) could be detected with wheat. Similar results were obtained in Scotland and Belgium (11,16). In comparison in inoculated Setaria italica activities at booting were up to 3000 nmoles/h/g dry weight (13). However, only very small amounts of $^{15}N_2$ were incorporated into the plant (13). In experiments carried out under different environmental conditions with different wheat cultivars, acetylene reduction was followed in cores during plant ontogeny. Low activities ranging from 30 to 80 nmoles ethylene/h/g dry weight, could be measured at 20°C only near heading and flowering stages of wheat. No acetylene reduction was detected in early stages of wheat ontogeny (Y. Kapulnik, unpublished).

Effect on root development and mineral uptake

The roots of wheat were sampled at different times during growth from field experiments carried out in winter of 1981-1982 in Israel (9). In all samples, inoculated wheat roots were far more developed than the roots of the controls. This occurred at three different levels (0, 60,120 Kg/ha) of initial nitrogen fertilizer. Similar effects were observed in greenhouse experiments in pots with soil and in hydroponics systems without detectable nitrogen fixation activity (9). Enhanced development of roots has also been observed in inoculated pearl millet (19) and in Setaria italica (7).

It can therefore be concluded that colonization of roots by Azospirillum clearly enhances root branching and development, starting at early stages of plant growth.

Increased accumulation of dry matter and of N, P and K in the shoots of Sorghum (17) and of wheat (9) has been observed in field experiments in Israel. Also in hydroponics systems, we have obtained an increase in the rates of NO_3^- uptake from the solution by inoculated roots (Y. Kapulnik, unpublished data). Similarly, it was reported by Lin et al. 1983 that in inoculated 3 days and 2-week old root segments of corn, there was an enhancement of 30 to 50% over controls in the uptake of NO_3^-, K^+ and H_2PO_4 from the solution. The increased mineral accumulation and uptake observed, could be due to either root size and branching or to increase in the absorption efficiency per unit of root. This remains to be demonstrated.

It seems therefore that Azospirillum can benefit plant growth by enhancing mineral uptake by the roots starting just after germination.

In order to find the effect of Azospirillum inoculation on germination and on early root development in wheat, five surface sterilized wheat seeds (T. aestivum L. cv Miriam) were germinated on filter paper in a petri dish and were inoculated 72 h later by 1 ml suspension containing a mixture of several Azospirillum strains. Total root length was measured after incubation for 7 days at 20°C. It was found in a preliminary experiment that inoculum consisting of washed cells (in tap water) significantly enhanced root elongation as compared to uninoculated controls. However, inoculation with cells resuspended in the growth medium (malate liquid medium) caused a decrease in root elongation. This suggested that azospirilla that were able to proliferate in the medium, caused a negative effect.

Subsequently germinated surface sterilized seeds were inoculated with a mixture of Azospirillum brasilense Cd, Sp 7 and the local isolate Cd-1 at a final concentration of 10^2-10^{10} colony forming units (CFU). Pots with vermiculite were inoculated by mixing the bacteria to a final concentration of 10^2-10^9 CFU per g of vermiculite. After incubation, 7 days for petri dishes and 20 days for vermiculite grown plants, the roots were carefully washed and the root length was measured. Total surface of the root system was determined by the titration method and by gravimetric method (3).

It was found that 10^5–10^6 CFU were optimal for enhancing root elongation and total surface, whereas 10^8–10^{10} CFU caused inhibition of root development (Fig. 1). Higher inoculum concentrations were needed to exert the same effects in seedlings incubated under lower temperatures.

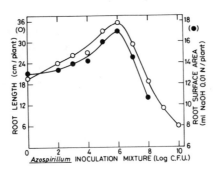

Fig. 1: The effect of Azospirillum concentration in the inoculum on root length (Petri dishes) and on root surface area (pots with vermiculite).

Seven different modern cultivars of spring wheat (Triticum aestivum var. aestivum L. emend. Thell) were tested to estimate their reaction to inoculation with an Azospirillum mixture 10^6 CFU/ml. It was found that inoculation increased root elongation above controls but there were no marked differences between the tested cultivars. Based on experiments carried out under controlled environmental conditions, it has been suggested that there are some specific interactions between different wheat species and cultivars and the type of bacterial isolates used for inoculum, thus suggesting that genotype may be important (14,15,16). This subject remains to be demonstrated by further experimentation in the field.

Azospirillum was compared to several types of bacteria (Table 2). Only Azospirillum (10^6 CFU/g vermiculite) caused a significant effect in enhancing wheat root surface area. In petri dishes both Azospirillum and Azotobacter enhanced root elongation (Table 2).

Table 2: Comparison of _Azospirillum_ with other bacteria on their effect on root development

	Bacterial conc. CFU	Petri dishes		Pots with vermiculite	
		Root length cm/plant	% increase	Root surface area mg $Ca(NO_3)_2$/plant	% increase
Klebsiella penumoniae	10^6 10^8	31.0±0.4 31.0±0.4	6.8 (NS) 6.8 (NS)	97.0±9.0 95.2±7.9	18.2 (NS) 15.8 (NS)
Azotobacter chroococcum	10^6 10^8	37.0±0.7 35.0±0.5	27.6 * 20.6 *	78.7±4.9 78.7±1.5	-4.0 (NS) -4.0 (NS)
Azospirillum sp.	10^6 10^8	35.0±0.6 29.4±0.4	20.6 * 1.3 (NS)	109.7±5.5 83.7±2.2	33.7 * 2.1 (NS)
Bacillus megaterium	10^6 10^8	33.0±0.6 29.8±0.6	13.8 (NS) 2.7 (NS)	77.5±3.7 77.2±2.0	-5.4 (NS) -5.8 (NS)
Bacillus subtilis	10^6 10^8	22.9±1.2 30.9±0.6	-21.0 (NS) 6.6 (NS)	87.5±4.5 88.7±3.5	6.7 (NS) 8.1 (NS)
Pseudomonas syringae pv tomato	10^6 10^8	31.0±0.5 25.8±0.6	6.9 (NS) -11.0 (NS)	87.5±2.2 90.5±2.5	6.7 (NS) 10.3 (NS)
CONTROL (dead Azospirillum cells)	0	29.0±0.8	−	82.0±4.5	−

± S.E. of the mean
* significance at level of 0.05 by using T test (df = 19)
NS Not significant
Root surface area calculated by the gravimetric method (3).

In preliminary studies (Y. Kapulnik, unpublished), scanning electron micrographs of inoculated wheat root segments showed denser and longer root hairs as compared to the controls treated with dead cells. In inoculated roots bacteria were located mainly on the cell elongation zone and on the basis of root hairs, but much less bacterial cells were present on the root cap or absorbed to root hairs.

Conclusions

The data obtained so far in Israel, in spring wheat inoculated with Azospirillum, indicated that after germination, the emerging roots are colonized, their growth and branching is promoted, mineral uptake from the soil solution is increased, thus promoting accumulation of dry matter and of minerals.

It remains to be demonstrated the colonization dynamics of roots by Azospirillum. Does it occur only at the beginning of growth or does it continue during plant ontogeny? Is the enhancement of root elongation and branching caused by growth promoting substances produced by the introduced organisms or is it a reaction of the roots to the infection? Are there any other factors such as protection of the roots from damaging pathogens or synergistic effects. Moreover, the partial contribution of nitrogen fixation on yield remains to be properly assessed mainly in the field.

Acknowledgements

This research was supported by a grant from the United States-Israel Binational Foundation (BSF), Jerusalem, Israel (grant No. 2476/81) and by the USA-Israel Binational Agriculture Research and Development Fund (BARD), Grant No. I-254-80.

References

1. van Berkum, P. and Bohlool, B.B. 1980. Microbiol. Rev. 44, 491-517.
2. Boddey, R.M. and Döbereiner, J. 1982. 12th Congress of Soil Science, New Delhi, India. pp. 28 -47.
3. Carley, H.E. and Watson, R.D. 1966. Soil Sci. 102, 289-291.
4. Eskew, D.L., Eaglesham, A.R.J. and App, A.A. 1981. Plant Physiol. 68,

48 -52.

5. Hegazi, N.A., Khawas, H. and Monib, M. 1981. Current Perspectives in Nitrogen Fixation. Gibson, A.H. and Newton, W.E. (Eds.). Australian Academy of Science, Canberra pp. 453.

6. Kapulnik, Y., Sarig, S., Nur, I., Okon, Y., Kigel, J. and Henis, Y. 1981. Exp. Agric. 17, 179-187.

7. Kapulnik, Y., Okon, Y., Kigel, J., Nur, I. and Henis, Y. 1981. Plant Physiol. 68, 340-343.

8. Kapulnik, Y., Sarig, S., Nur, I. and Okon, Y. 1983. Can. J. Microbiol. In press.

9. Kapulnik, Y., Sarig, S. and Okon, Y. 1982. Second Int. Symp. in N_2 Fixation with Non-legumes. Banff, Canada 5-10 September 1982, p. 26.

10. Lin, W., Okon, Y. and Hardy, R.W.F. 1983. Appl. Environ. Microbiol. 45, 1775 - 1779.

11. Lethbridge, G., Davidson, M.S. and Sparling, G.P. 1982. Soil Biol. Biochem. 14, 27-35.

12. Okon, Y. 1982. Isr. J. Bot. 31, 214-220.

13. Okon, Y., Heytler, P.G. and Hardy, R.W.F. 1983. Appl. Environ. Microbiol. In press.

14. Rennie, R.J. and Larson, R.I. 1979. Can. J. Bot. 57, 2771-2775.

15. Rennie, R.J., De Freitas, J.R., Ruschel, A.P. and Vose, P.V. 1983. Can. J. Bot. 61, in press.

16. Reynders, L. and Vlassak, K. 1982. Plant Soil 66, 217-223.

17. Sarig, S., Kapulnik, Y., Nur, I. and Okon, Y. 1984. Exp. Agric. 20, in press.

18. Schank, S.C., Weier, K.L. and MacRae, I.C. 1981. Appl. Environ. Microbiol. 41, 342-345

19. Tien, T.M., Gaskins, M.H. and Hubbell, D.H. 1979. Appl. Environ. Microbiol. 37, 1016-1024.

20. Vlassak, K. and Reynders, L. 1982. Vol.1 Associative N_2 Fixation. Vose, P.B. and Ruschel, A.P. (Eds.) CRC Press, Inc. Florida, pp. 93-101.

CONTRIBUTION OF AZOSPIRILLUM SPP. TO ASYMBIOTIC N_2-FIXATION
IN SOILS AND ON ROOTS OF PLANTS GROWN IN EGYPT

N.A. Hegazi

Department of Microbiology, Faculty of Agriculture,
Cairo University, Giza, Egypt

Introduction: Some aspects of asymbiotic N_2-fixation in Nile Valley soils

Soils of Nile Valley are considered one of the unique environments where asymbiotic N_2-fixation contribute significantly to the N-status in soils and plants (1, 23, 25, 32). Evidences presented were both based on Kjeldahl analysis (1, 12) and acetylene-reducing activity (23). Of particular interest are the results of in-situ measurements of acetylene reduction recorded in Giza fields just post harvest of winter (wheat) and summer (maize) crops (27). The enrichment of such soils with significant amounts of straw and plant refuses-amounted to ca 4 tons ha^{-1}-accelerated growth and activities of both heterotrophic bacteria and blue-green algae. Rates of acetylene-reducing activity followed a distinguished pattern throughout the duration period of each of the successive flooding irrigation cycles (Table 1). They were rather high at wet conditions (moisture content of >50-100% W.H.C.) prevailed during the first few days after completion of flooding irrigation. This was followed by remarkable decreases in activities measured. A conclusion was reached that asymbiotic N_2-fixation in soil is a factor of available organic matter of wide C/N ratio and of moisture content (Table 2). In summer, high temperature (>25-32°C), quick dryness and dessication of soil (<20% moisture content) within a week after flooding do not support full growth of diazotrophs particularly blue-green alage. Calculated amounts of N_2-fixed corresponded to ⩽40 kg N ha^{-1}. Modification of the environment in winter time exerted stimulatory effects on diazotrophs when plant refuses of previous standing crops are incorporated into soil as well. Flooding and relative slow dryness of the soil, together with temperature in the range of >14-22°C

Table 1: In-situ acetylene-reducing activity measured in Giza fields (n mol C_2H_4 kg^{-1} day^{-1}) during summer and winter post organic matter incorporation and flood irrigation

Days after irrigation	February (14°C)			July (28-32°C)		
	Moisture %	Acetylene reduction in		Moisture %	Acetylene reduction in	
		Transparent bottles (a)	Brown bottles (b)		Transparent bottles (a)	Brown bottles (b)
1	26.5	17.90	0.38	35.5	1.22	0.86
2	25.5	29.62	2.33	33.5	4.56	3.70
3	23.5	44.98	4.34	29.5	5.88	5.93
4	22.9	46.70	10.97	27.3	4.56	6.50
5	22.6	43.44	16.46	24.4	3.38	7.87
6	22.2	39.07	20.33	21.9	1.44	7.22
7	18.9	31.75	17.06	21.2	0.07	3.74
8	18.2	10.30	10.87	19.4	0.05	1.92
9	ND	ND	ND	17.8	0.05	0.86
10	ND	ND	ND	15.4	0.02	0.36
11	ND	ND	ND	10.9	0	0.07
12	17.8	7.46	3.84	10.5	0	0.05

- Based on data presented by Khawas[27]
- Post harvest of wheat (winter) and maize (summer), straw and plant refuses incorporated into soil. 2-l bottomless transparent or brown flasks inserted into flooded soil to a depth of 10 cm. Outlets of flasks stopped with Suba-seal rubber closures. Every morning 10% of gas phase was replaced by C_2H_2 and C_2H_4 measured throughout the day.
- Statistical analysis indicated significant difference (P<0.01) between a and b, a and c, b and d; no significant differences between c and d.

activated growth and N_2-fixation (20.33 - 46.70 µ mol C_2H_4 kg^{-1} day^{-1}) of diazotrophs mainly blue-green algae (Table 1).

Functional groups of diazotrophs

A heterogenous community of diazotrophs is actively involved in asymbiotic N_2-fixation in Nile Valley soils. Azotobacter (1, 12) occur with higher population densities than in soils of neighbouring countries (12, 13, 16). Besides, Azospirillum spp (14, 15, 17), Clostridium spp (1) and N_2-fixing Bacillus spp and Gram-negative short rods of genera Klebsiella spp. and Enterobacter spp. (17) are common. Representative of the latter groups of diazotrophs - compared to Azotobacter - are characterized by their dynamic response to envrionmental changes. For example Table (2) indicates their fast growth and substantial increases in their population densities in response to combined effects of incorporation of plant refuses in soil and waterflooding (27). Decreases in numbers of diazotrophs attributed to exhaustion of organic matter and dryness of soil were highest for Azospirilla and lowest for azotobacters.

Table 2: Changes in population densities of diazotrophs in response to combined effect of organic matter[a] incorporation into soil and flooding irrigation during summer time (July)

Diazotrophs	No. of bacteria ($\times 10^3 g^{-1}$ soil)		Rates of increase (fold)
	Before irrigation and organic matter incorporation	Within a week post irrigation & organic matter incorporation	
Azotobacter	0.4	14.9	37.3
Azospirilla	1.5	5330.0	3553.3
N_2-fixing rods	2.3	3700.0	1608.7
Clostridia	34.7	11635.0	335.3
Oligonitrophs	119.0	11530.0	96.9
Moisture content (%)	4.9	23.3	
Total organic carbon (%)	0.99	1.07	

Based on data present by Khawas[27]

(a) Straw and plant refuses applied with a rate of ca 4 tons ha^{-1}

Ecology of Azospirillum spp

Azospirillum spp. are of ubiquitous distribution in Egyptian soils. They are consistently isolated from enrichments prepared for soils of various localities and textures, and under various standing crops (2, 3, 5, 6). Their densities exceed those of other diazatrophs, particularly Azotobacter (Table 3). They are also reported common in water and sediment of River Nile as well as on roots of water hyacinth (24) and papyrus plants (Higazy, personal communication).

Table 3: Distribution of Azospirillum spp compared to Azotobacter spp. in Egyptian soils.

Densities	Nile Valley[a]		Outside Valley[b]	
	Azospirilla	Azotobacter	Azospirilla	Azotobacter
<10	-	-	-	3
$>10-10^2$	-	-	3	3
$>10^2-10^3$	1	14	3	-
$>10^3-10^4$	15	7	1	1
$>10^4-10^5$	7	3	-	-
$>10^5-10^6$	1	-	-	-
Total	24	24	7	7

- Based on data presented by Hegazi et al.[15], Hegazi et al.[20], Khawas[27] and Makboul et al.[28]

(a) Samples obtained from Upper and Lower Egypt.
(b) Samples obtained from Sinai, Northern coast and Wadi-Hoff in Western Desert.

Azospirillum spp. are associated to roots of plants - belonging to various families - grown in Egypt (3, 6). They are enriched to a great extent in the rhizosphere of major cereals and are present on their root surfaces (27). This particular group of diazotrophs as well as N_2-fixing bacilli and Gram-negative short rods have got a peculiar affinity to roots of maize and sorghum. On the other hand, Azotobacter have got a unique distribution being found with appreciable numbers only on root surfaces of wheat (Table 4). This might support the use of Azotobacter as an inoculum for wheat and explain positive results of inoculation trials (11, 29). Azospirillum spp were also found among microflora of phylloplanes of cereals (30). They are reported to colonize - better than Azotobacter - roots of desert plants widely distributed in Egyptian Western Desert (20, 27, 33).

Table 4: Counts of diazotrophs ($X10^4 g^{-1}$) recorded in the root region of 3-4 months' old grasses grown in Giza fields

Plants	Rhizosphere	Ectorhizosphere	Endorhizosphere (a)
		Azotobacter (colony counts)	
Zea maize, L.	6.2	40.0	0
Sorghum vulgare, Pers.	9.2	36.0	0
S.vulgare var. Saccharatum, Boerl.	6.6	36.0	0
S.vulgare var. technicum, Jav.	8.8	15.0	0
Triticum aestivum, L.	184.0	380.0	3.2
Hordeum vulgare, L.	10.2	46.0	0
P.turgidium [b]	0	0	0
		Azospirillum (m.p.n.)	
Zea maize, L.	11.0	650.0	4.3
Sorghum vulgare, Pers.	110.0	7100.0	120.0
S.vulgare var. Saccharatum, Boerl.	41.0	70.0	3.6
S.vulgare var. technicum, Jav.	9.0	900.0	14.0
Triticum aestivum, L.	42.0	200.0	11.0
Hordeum vulgare, L.	15.0	260.0	4.7
P.turgidium [b]	2.3	130.0	1.8

- Based on data present by Khawas [28]
a, recently termed as histosphere.
b, a common desert plant, belongs to Gramineae and prevails in Egyptian Western Desert.

Characteristics of prevailing types of Azospirillum spp.

Several investigations included isolation, characterization and identification of representative types of Azospirillum spp. prevailing in various terrestrial and aquatic

habitats (2, 3, 5, 6, 24, 26, 27). Strains obtained were placed
in either species of A.lipoferum or A.brasilense. However, the
taxonomic status of quite a number of isolates is remained un-
certain as they showed peculiar cell morphology, motility and
flagellation when grown in either N-deficient or N-enriched
media (15, 22). Besides to various characteristics studied (9,
26), pattern of accumulation of poly-β-hydroxybutyrate in cells
of A.lipoferum was established (3, 4). The synthesis of such
polymer followed closely bacterial growth curve and reached
climax - as it represented ca. 30% of cell dry weight - by the on-
set of the stationary phase of growth. Possible utilization of
the stored polymer by cells of Azospirilla under starving conditions was demonstrated as pure strains were able to utilize
the compound as efficient as glucose and other favourable organic
acids (4, 17).

Inoculation trials

Pot experiments

Hegazi et al. (21) first tried inoculation of maize
seedlings - grown under subtropic conditions provided in a greenhouse - with dense suspensions of either A.vinelandii or A.lipo-
ferum (ATCC 29145). Performance of Azospirilla - but not Azotobacter - inocula was much more pronounced on roots of plants
grown in Nile Valley soils compared to fertile alluvial Belgian
soils. Population densities and N_2-ase activity of Azospirilla
were significantly increased and amounts of N_2-fixed was doubled
through inoculation.

Further experimentation with maize grown under Egyptian
conditions were undertaken (19). Inoculation with mixed inocula
of A.lipoferum significantly increased total dry weight and
N-content (Table 5).This was correlated with significant increases
in numbers of Azospirilla as well as of other various groups
of soil microorganisms as indicated by ATP determinations. The
effect of straw amendment (2%, w/w) on establishment of azospirilla inocula was studied, and it was found that spontaneous
azospirilla inoculation and straw addition exerted the most

Table 5: Effect of inoculation with Azospirilla on agronomic and biological parameters of 12 weeks-old maize plants[a] grown in presence or absence of straw amendment.

Parameters	Untreated	Azospirilla inoculated	Soil + 2% straw	Azospirilla inoculated + 2% straw	L.S.D. (0.05)
Dry weight (g/plant)	11.4	22.9	20.1	39.1	8.5
Leaf surface area (cm^2 $plant^{-1}$)	232	396	364	1110	54.0
Plant height (cm)	64	79	81	108	14.3
N-content (%)	0.71	1.12	1.06	1.39	0.15
m.p.n. estimates of azospirilla in histosphere ($x10^4$ g^{-1} dry matter)	8.1	940.0	8.2	1260.0	32.5
N_2-ase activity on roots (n mol C_2H_4 h^{-1} g^{-1})	26.3	114.1	102.7	144.7	18.6
ATP in rhizosphere soil (ng ATP g dry $root^{-1}$)	96.0	473.0	394.0	521.0	87.0
Total organic carbon in rhizosphere soil (%)	0.93	0.84	1.62	1.84	0.27

Based on data presented by Hegazi et al. (19)

a. Seedlings of the local variety American Early were soaked in dense mixed broth of A.lipoferum originally isolated from maizes grown in Egypt. Plants grown in 10-l pots containing Nile clay soil and kept under greenhouse conditions (30000 lx; 16-h day of 28-30°C; 8 h night of 18-22°C).
b. Maize straw (C/N ratio of 54.9:1) added to soil (2% w/w).

distinguishable positive effects on development of maize plants. Components of straw and/or its degradative compounds, e.g. organic acids, furnish dense populations of diazotrophs - including introduced Azospirilla - with significant amounts of carbon and energy sources. This particular treatment did boost development of sorghum plants as well (10) and significant increases in dry weight and N-content of plants, particularly in leaves, were obtained. Positive correlation was recorded among ATP concentrations, population densities of Azospirilla and N_2-ase activity on roots. Values of correlation coefficients decreased in absence of azospirilla inocula and in presence of straw (5% w/w) which indicated that straw stimulated - besides azospirilla - various groups of diazotrophs. Inoculation of maize as well as sorghum seedlings with azospirilla as such did expand the ecological niche of these particular diazotrophs in the root region. Their numbers increased many folds in the histosphere, particularly when inoculation combined with straw amendment. It is indicated from data presented by Hegazi et al. (19) and Eid et al. (10) that maximum numbers of azospirilla and rates of N_2-ase activity on root surfaces, as the active sites of associative N_2-fixation, are reported during the flowering stage of sorghum plants. A phenomenon that was already reported for wheat plants (Hegazi, unpublished data).

Using the dilution ^{15}N-technique, Eid (6) and Eid et al. (7) were able to demonstrate active N_2-fixation - which accounted for >10-47% of N-content of plant - associated with roots of sorghum plants inoculated with A.lipoferum (Table 6).

Table 6: Effect of various doses of ^{15}N-fertilization on N_2-fixation on roots of inoculated sorghum plants as measured with ^{15}N-dilution technique

Age of plant (days)	Dose of $^{15}NH_4^+$-N (mM)	Total-N content (mg/plant)	N-content of plant derived from			
			N-fertilizer (mg/plant)	(%)	N_2-fixation (mg/plant)	(%)
25	0.52	29.49	12.14	41.2	2.98	10.1
	0.26	31.90	8.27	25.9	9.35	29.3
	0.13	25.84	5.94	23.0	5.53	21.4
55	0.52	43.30	14.90	34.4	14.18	32.7
	0.26	43.50	8.49	19.5	20.63	47.4
	0.13	37.27	5.38	14.4	17.52	47.0

Based on data presented by Eid (6) and Eid et al. (7).
Sorghum seedlings inoculated with mixed broth cultures of A.lipoferum then grown in 3-1 pots containing fine washed sand. $(^{15}NH_4)_2 SO_4$ (32.86 ^{15}N atom%) added to sand culture in various doses (0.53, 0.26 and 0.13 mM $^{15}NH_4^+$). Greenhouse conditions were: 12-h day (30000 lx; 30-36°C); 12-h night (28-30°C). The isotopic abundance of the nitrogen found in plant samples obtained at various stages of growth was liberated by Dumas method and measured by optic emission analysis.

Substantial decreases in ^{15}N- abundance were accomplished by the successive growth of plants and by decreasing initial concentrations of $^{15}NH_4^+$-N added to the washed sand cultures. Similar to acetylene-reduction assay, the isotope dilution technique recorded increasing rates of N_2-fixation up to the grain filling stage. The C_2H_2/N_2 molar ratios calculated from slopes fall near to the theoretical one; being 3.94:1, 4.08:1 and 3.13:1 under the soil environments contained 0.53, 0.26 and 0.13 M $^{15}NH_4^+$-N respectively. Applying the same technique, Amer (3) presented evidences for active N_2-fixation associated with roots of maize plants inoculated with A.lipoferum. He estimated as much as 45% of total N-content of plant derived from N_2-fixation.

Hegazi (unpublished data) investigated the response of Egyptian wheat variety Giza 156 to inoculation with strains of A.brasilense of different origin (Tables 7, 8, 9 and 10). The strain of azospirilla isolated from roots of German grasses led to pronounced increases in dry weight as well as other parameters measured (Table 7). Increases of total dry weight (Table 8) attributed to spontaneous azospirilla inoculation and application of moderate dose of N (0.5 g N kg^{-1} soil) were particularly higher than those reported in presence of full N-dose (1.0 g N kg^{-1} soil). The inoculation process as such stimulated developments of roots as an increase of >15-40% in root dry weight was reported over untreated plants (Table 8). Application of moderate dose of N-fertilizers together with inoculation reduced such increase to <10%, and of full N-dose significantly inhibited root growth. Population densities of azospirilla - but not N_2-ase activity and N-content of plants - were significantly increased due to inoculation particularly in presence of moderate dose of N (Table 9). Plants of the latter treatment developed vigorously, being taller and contained higher amounts of chlorophyl than those receiving full doses of N (Table 10). This might be attributed to better root development of plants of this particular treatment led to efficient nutrients' uptake, and to general improvement of physiological status of inoculted plants as due to possible production of growth regulators by azospirilla.

Table 7: Response of Egyptian wheat variety Giza 156 to inoculation with A.brasilense of various origin [a]

Parameter	Egypt (strain Eg W_2)		West Germany (Wg G_1)	
- m.p.n. estimates of azospirilla in histosphere ($\times 10^6$ g dry root^{-1})	12.3	(86.7%)[b]	33.0	(266.7%)
- N_2-ase activity on roots (n mol C_2H_4 h^{-1} g dry root^{-1})	177.0	(402.0%)	307.0	(772.0%)
- Dry weight (g/pot)				
Total	3.83	(2.6%)	4.06	(8.6%)
Ears	1.10	(0.3%)	1.20	(9.0%)
Roots	0.83	(15.2%)	0.99	(37.4%)
- Plant height (cm)	36.9	(12.5%)	35.7	(8.8%)
- Chlorophy content (μg g fresh weight^{-1})				
chlorophyl a	216.79	(0.0%)	254.18	(15.4%)
chlorophyl b	65.88	(75.8%)	73.32	(95.7%)

a : Seeds were grown in 1 kg pot containing mixture of Nile clay soil and sand (1:1, w/w, pH 7.6). Seven treatments were prepared: untreated; sprayed several times during vegetative growth with growth regulators (mixture of 30 ppm of Kinetine, indol acetic acid and gibberellic acid); N-fertilized (1.0 g N kg^{-1} in the form of $(NH_4)_2 SO_4$ in 2 equal doses); azospirilla inoculated + 1/2 dose (0.5 g n Kg^{-1}) N-fertilized plants.

16 seeds were sown in each pot after being soaked in broth cultures as such (for inoculated treatments) or previously autoclaved (for other treatments). Later, 1 ml broth added to every sown seed. 0.65 g KH_2PO_4 was added for each pot of all treatments just after germination. Growth conditions in greenhouse were 13-h day (15000 lx; 20-28°C) and 11-h night (15-18°C).

b: Percentage increase over untreated plants.

Table 8: Effect of azospirilla inoculation and N-fertilization on development of wheat plants.

Treatments	Dry weight (g/pot)		
	Total plant	Ears	Roots
Untreated plants	3.73	1.10	0.72
Plants sprayed with regulators	3.72	0.48	0.65
Inoculated plants (Eg W_2)	3.83	1.10	0.83
Inoculated plants (Wg G_1)	4.06	1.20	0.99
Inoculated (Eg W_2) + 1/2 N dose	6.51	1.89	0.65
Inoculated (Wg G_1) + 1/2 N dose	6.27	1.87	0.69
Full N-fertilized plants	5.90	1.73	0.59

For more information on the experiment refer to legend of Table 7.

Table 9: Effect of azospirilla inoculation and N-fertilization on N_2-fixation associated with roots of wheat plants[a]

Treatments	m.p.n estimates of azospirilla ($\times 10^6$ g^{-1} dry root)	N_2-ase activity (n mol C_2H_4 h^{-1} plant^{-1})
Untreated plants	9.0	25.2
Plants sprayed with regulators	14.0	12.9
Full N-fertilized plants	15.0	16.5
Inoculated plants (Eg W_2)	12.3	177.0
Inoculated plants (Wg G_1)	33.0	307.6
Inocualted (Eg W_2) + 1/2 N dose	399.2	39.2
Inoculated (Wg G_1) + 1/2 N dose	389.7	44.2

For more information on the experiment refer to legend of Table 7.
a, Plants at 50% flowering stage of growth.

Table 10: Effect of azospirilla inoculation and N-fertilization on certain physiological parameters of wheat plants[a]

Treatment	Plant height (cm)	Chlorophyl content ($\mu g\ g^{-1}$ fresh weight)		N-content (%)
		a	b	
Untreated	32.8	220.2	37.5	0.43
Plants sprayed with regulators	49.0	196.6	30.4	0.57
Full N-fertilized plants	45.0	216.8	65.9	1.77
Inoculated plants (Eg W_2)	36.9	254.2	73.3	0.41
Inoculated plants (Wg G_1)	35.7	588.5	156.3	0.41
Inoculated plants (Eg W_2) + 1/2 N dose	49.0	ND	ND	1.19
Inoculated plants (Wg G_1) + 1/2 N dose	50.2	509.9	109.9	1.11

For more information on the experiment refer to legend of Table 7.
a, Plants at 50% flowering stage of growth.
ND, not determined.

Field trials

Several trials were executed in the experimental fields of Cairo University at Giza Governorate. As a general practice, mixed liquid inocula prepared by mixing equal volumes of separately-grown broth cultures of 3 potent isolates of either A. lipoferum (isolated from roots of Egyptian maize) or A. brasilense (isolated from roots of Egyptian wheat) were used for soaking of seeds just prior sowing. Responses to inoculation under natural field conditions prevailing in Egypt were much more pronounced than those reported for pot experiments carried out under greenhouse conditions with supplementary lighting and controlled day and night temperature. A number of factors are found to govern the outcome of azospirilla inoculation when applied as an agronomic practice. First of all, a significant genotype-inoculation interaction was reported. Associative N_2-fixation on roots and yields of tall but not short varieties of sorghum were particularly improved through azospirilla inoculation (8, Tables 11 and 12). This was related to a wide specific green area of the tall varieties that are available for intense photosynthetic metabolism which results in excess portions of photosynthetates translocated downwards to the site of associative N_2-fixation. Similarly, the significant improvement of yields of open pollinated - but not Double cross - varieties of maize by azospirilla inoculation is attributed most probably to their bigger specific green area, measured as leaf area index and plant height (19, Table 13). A phenomenon that was also reported among wheat varieities (18, 27, 31). The local tall variety Giza 156 responded more than the semi-dwarf Shnab 70 to inoculation with A.brasilense (Table 13).

Table 11: Population densities of azospirilla and N_2-ase activity on roots of 90 days-old sorghum plants grown under Giza field conditions.

Variety	m.p.n. estimates of azospirilla (x10^5 g^{-1})		N_2-ase activity (a) (n mol C_2H_4 h^{-1} g^{-1})
	Rhizo-sphere	Histo-sphere	
Giza 3 (Short variety)			
Untreated plants	2.6	3.5	94.3
+ 200 kg N ha^{-1}	1.8	2.0	55.7
Inoculated plants	14.3	41.9	338.2
Inoculated+50 kg N ha^{-1}	6.6	25.1	486.1
Giza 15 (Tall variety)			
Untreated plants	3.7	4.6	81.5
+ 200 kg N ha^{-1}	2.9	3.3	28.5
Inoculated plants	8.6	13.0	560.4
Inoculated+50 kg N ha^{-1}	11.0	62.1	608.0
Giza 114 (Tall variety)			
Untreated plants	0.4	5.6	122.1
+ 200 kg N ha^{-1}	4.4	5.4	49.4
Inoculated plants	2.9	16.2	892.8
Inoculated+50 kg N ha^{-1}	9.5	61.9	391.3

Based on data presented by Eid et al. (8)
a: n mol C_2H_4 evolved after 12-hr aerobic incubation with C_2H_2.

Available stock of soil nitrogen interact with the response of plant growth to azospirilla inoculation. It is reported that low or moderate levels of nitrogen (<50 kg N ha^{-1}) significantly intensified associative N_2-fixation (Table 11) and increased the yield components of sorghum plants inoculated with A.lipoferum (5, 8, Table 12).

Table 12: Effect of inoculation with azospirilla on yields of sorghum plants grown under Giza field conditions

Treatments	Varieties of Sorghum		
	V. Giza 3	V. Giza 15	V. Giza 114
Untreated plants	1194^a	1570	1327
	6263^b	6453	7493
	1.71^c	1.7	1.82
N-fertilized plants (200 kg N ha^{-1})	1918	3342	2710
	10038	12951	15488
	3.19	2.86	3.01
Inoculated plants	1918	2812	2780
	11758	16342	14254
	1.94	1.74	1.93
Inoculated+50 kg N ha^{-1} fertilized plants	2072	3747	3565
	12211	20153	17030
	2.72	1.90	2.19

Based on data presented by Eid et al. (8).
a : grain yield; b: straw yield; c: N(%)
L.S.D (0.05): for a grain yield, 370; for straw yield, 876; for N, 0.80.

Moisture content of the soil had a remarkable influence on maize and sorghum-azospirilla associations under Egyptian semi-arid conditions (3, 6, 8, 19). Increasing moisture content in soil up to field capacity by flood irrigation maximized N_2-ase activity on plant roots within the first three days post completion of irrigation. During this particular period, soil tended towards microaerophilic side which favours activity of diazotrops. Continuous drain of water and successive dryness of soil spread aerobiosis in the vicinity of roots resulting in further decreasing of N_2-ase activity until the next flooding irrigation is applied (Table 14). Therefore it is believed that flood-irrigation applied in the form of successive cycles, do modify the semi-arid conditions prevailing in Egypt and make the soil/root environment - during at least 1/4th of the duration of each cycle of irrigation - is favourable to asymbiotic N_2-fixation.

Table 13: Response of various parameters of maize plants to inoculation with azospirilla under Giza field conditions

	Varieties of Maize			Treatment means
	V.Cairo 1	Double Cross	American Early	
N_2-ase activity (n mol C_2H_4 h^{-1} g dry root^{-1})				
N-fertilized plants	12.8	6.5	10.1	9.8
Inoculated plants	92.4	36.9	66.5	65.3
Untreated plants	29.9	17.1	30.8	25.9
Variety means	45.0	20.1	35.8	-
Plant height (cm)				
N-fertilized plants	208	204	208	207
Inoculated plants	192	150	199	180
Untreated plants	147	141	144	144
Variety means	182	165	184	-
Leaf area index				
N-fertilized plants	6.3	6.0	6.2	6.2
Inoculated plants	5.2	2.5	5.3	4.3
Untreated plants	2.1	1.7	1.5	1.7
Variety means	4.5	3.4	4.3	-
Grain yield (kg ha^{-1}) and N-content (%)				
N-fertilized plants	8698(1.82)	3237(1.75)	8247(1.47)	8394(1.68)
Inoculated plants	8171(1.75)	3688(1.34)	4829(1.68)	5562(1.59)
Untreated plants	3218(1.42)	3347(1.13)	2648(1.28)	3070(1.28)
Variety means	6694(1.66)	5091(1.43)	5241(1.48)	- -

Based on data presented by Hegazi et al. (19).
L.S.D. (0.05) for treatments (32.5, 6.5, 0.6, 263.9 and 0.22) and for varieties (15.2, 10.2, 0.4, 230.6 and n.s.) were calculated for N_2-ase activity, plant height, leaf area index, grain yield and N-content respectively.

Table 14: Effect of inoculation with A.brasilense on growth of different varieties of wheat plants grown under field conditions

Parameters	Variety Shnab 70			Variety Giza 156		
	Un-treated	N-fertilized	Inoc-ulated	Un-treated	N-fertilized	Inoc-ulated
m.p.n. estimates of azospirilla [a]						
in rhizosphere	1.5	0.8	675.0	1.3	0.9	2830.0
in histosphere	35.0	4.3	60.0	37.0	4.3	102.0
R/S value	3.0	1.6	1350.0	2.6	1.8	5660.0
N_2-ase activity [b]						
8 hrs	2.8	3.5	7.3	3.7	2.6	7.2
24 hrs	1.2	3.0	14.5	4.7	1.0	10.6
Plant height (cm)	64.1	68.7	77.1	71.2	78.6	90.1
No. of tellers per plant	1.7	1.6	3.1	1.7	2.6	3.0
Dry weight (g/plant)						
Shoot	2.39	6.72	7.47	4.03	6.98	8.76
Root	0.15	0.13	0.38	0.21	0.23	0.34
Yield (kg feddan^{-1}) [c]						
Grain	199	898	939	335	1007	1350
Straw	678	1470	1802	658	2240	2055
Protein content (%)	11.26	12.53	11.95	10.62	12.72	11.23

Base on data presented by Monib et al. (31)
a : $X10^4$ g^{-1} dry root of 2 months-old plants.
b : n mol C_2H_4 h^{-1} g^{-1} dry roots of 3 months-old plants after 8-24 hr aerobic incubation with C_2H_2
c : feddan = 4200 m^2

Table 15: Aerobic N_2-ase activity[a] measured on non-washed excised roots of 90 days-old sorghum plants throughout a single cycle of flooding irrigation[b]

Treatment	Days after completion of irrigation				
	1	2	4	7	9
Untreated plants	2300	2200	2100	70	13
Inoculated plants	3000	2900	2800	200	15
Inoculated plants + 50 kg N ha^{-1}	2600	2500	2400	80	10
N-fertilized plants (200 kg N ha^{-1})	90	80	75	10	5
Moisture (%)	54.8	45.8	43.8	32.4	27.2

Based on data presented by Eid et al. (8).

a : n mol C_2H_4 g^{-1} h^{-1} assayed after 4-8 hrs incubation with C_2H_2.

b : The field was flooded with Nile water. Roots of 10 plants were sampled and assayed for aerobic N_2-ase activity during this particular cycle of irrigation which covers a period of ca. 10 days.

References
1. Abd-el-Malek, Y. 1971. Plant and Soil. Special Volume 423-442.
2. Amer, H.A. 1978. M.Sc. Thesis. Fac. Agric., Cairo Univ.
3. Amer, H.A. 1982. Ph.D. Thesis. Fac. Agric., Cairo Univ.
4. Amer, H.A., Monib, M. and Hegazi, N.A. 1983. Egyptian Society of Applied Microbiol. Vth Conf. Microbiol. Cairo, May 1983 (in press).
5. Eid, M. 1978. M.Sc. Thesis, Fac. Agric. Cairo Univ.
6. Eid, M. 1982. Ph.D. Thesis. Fac. Agric. Cairo Univ.
7. Eid, M., Hegazi, N.A., Jensen, V. and Monib M. 1983. Egyptian Society of Applied Microbiology. Vth Conf. Microbiol. Cairo, May 1983 (in press).
8. Eid, M., Hegazi, N.A., Monib, M. and El-Sayed Shokr. 1983. Rev. Ecol. Biol. Sol (in press).
9. Eid, M., and Jensen, V. 1983. Egyptian Society of Applied Microbiology. Vth Conf. Microbiol. Cairo, May 1983 (in press)
10. Eid, M., Monib, M., Jensen, V. and Hegazi, N.A. 1983. Energy from Biomass. 2nd E.C. Conference. A. Strub, P. Chartier and G. Schleser, eds. Applied Science Publishers, Essex, England, pp. 205-209.
11. Fouad, M.F. 1981. Ph.D. Thesis. Katholieke Univ., Leuven, Belgium.
12. Hegazi, N.A. 1979. Zbl. Bakt. II. Abt. $\underline{134}$:489-497.
13. Hegazi, N.A., and Al-Sahael, Y. 1983. Egyptian Society of Applied Microbiology. Vth Conf. Microbiol. Cairo, May 1983 (in press).
14. Hegazi, N.A., Amer, H., and Monib, M. 1979. Soil Biol. Biochem. $\underline{11}$:437-438.
15. Hegazi, N.A., Amer, H.A. and Monib, M. 1980. Rev. Ecol. Biol. Sol. $\underline{17}$:491-499.
16. Hegazi, N.A., and Ayoub Saneya. 1979. Zbl. Bakt. Abt. $\underline{134}$:536-543.
17. Hegazi, N.A., Eid. M., Farag, R.S. and Monib, M. 1979. Rev. Ecol. Biol. Sol. $\underline{16}$:23-37.

18. Hegazi, N.A., Khawas, H. and Monib, M. 1981. Proceeding of IV Int. Sym. on Nitrogen Fixation. Canberra, Australia. Australian Academy of Science Publications, pp. 493.
19. Hegazi, N.A., Monib, M., Amer, H.A. and El-Sayed Shokr. 1983. Can. J. Microbiol. (in press).
20. Hegazi, N.A., El-Mallawani, A.A., and Monib, M. 1980. Egyptian Society of Applied Microbiology. Proceeding of IV Conf. Microbiol. Cairo. 1980, 119-123.
21. Hegazi, N.A., Monib, M. and Vlassak, K. 1979. Appl. Environ. Microbiol. 38:621-625.
22. Hegazi, N.A. and Vlassak, K. 1979. Folia Microbiol. 24:376-378.
23. Hegazi, N.A., Vlassak, K. and Monib, M. 1979. Plant and Soil, 51:27-37.
24. Higazy, A.M. 1980. M.Sc. Thesis. Fac. Agric., Cairo Univ.
25. Jensen, H.L. 1965. In "Soil Nitrogen". M.V. Bartholomew and F.E. Clark eds. Am. Soc. Agron. Mongr. 10:436-480.
26. Khalil, E.F. 1981. M.Sc. Thesis. Fac. Agric., Cairo Univ.
27. Khawas, H.M. 1981. M.Sc. Thesis. Fac. Agric., Cairo Univ.
28. Makboul, H.E., Fayez, M., Nadia F. Emam and El-Shahawy, R. 1983. Egyptian Society of Applied Microbiology. Vth Conf. Microbiol., Cairo, May 1983 (in press).
29. Monib, M., Abd-el-Malek, Y., Hosny, J. and Fouad, M. Fayez. 1979. Zbl. Bakt. II. Abt. 134:140-148.
30. Monib, M., Eid, M., Amer, H.A. and Hegazi, N.A. 1979. Egyptian Society of Applied Microbiology. Proceedings of Annual Meeting, Cairo 1979, pp. 141-158.
31. Monib, M., Hegazi, N.A., El-Sayed Shokr and Khawas, H.M. 1981. Research Bulletin of Fac. Agric., Ain Shams Univ. 1535:1-25.
32. Mulder, E.G. and Brotonegoro, S. 1974. In "The Biology of Nitrogen Fixation". A. Quispel, ed. North Holland-American Elsevier, pp. 37-85.
33. Othman, B.A.A. 1979. M.Sc. Thesis, Fac. Agric., Ain Sham Univ., Cairo.

RESUMÉ

In the last two years, research on Azospirillum showed considerable intensification as reflected by the increase in the number of participants of this workshop, compared to the first workshop in 1981.

Genetic studies showed progress in the collection of mutants of different types. Several nif structural mutants, a <u>nif</u> A type regulatory mutant, and mutants probably defective in general nitrogen control were obtained; in addition several resistance mutants, and such with alterations in a number of physiological or metabolic properties were reported. By molecular methods, a homology of the <u>nif</u>-HDK genes and the <u>nif</u>-A region of <u>Klebsiella pneumoniae</u> with DNA fragments of Azospirillum has been shown. The <u>nif</u>-HDK homologous region has been cloned and physically mapped. Mutants mentioned above may be helpful to elucidate, to what extent the observed inoculation effects are due to either nitrogen fixation or other factors.

Recent contributions towards a better understanding of the physiology and metabolism of Azospirillum, bearing on its possible association with plant roots, are rather scarce. Strong aerotaxis by Azospirillum has been demonstrated. Chemotaxis to individual sugars, organic acids, and root extracts has also been found, but is difficult to interpret in the light of the aerotaxis mentioned. Some of the enzymes related to the adaption of the organism to high and low pO_2 have been identified, and denitrification in <u>A. brasilense</u> Sp 7, associated with plants under anaerobic conditions was reported. An influence on root development of wheat, dependent on inoculum size of Azospirillum was observed for strain <u>A. brasilense</u> Cd.

A new species, <u>Azospirillum amazonense</u> has been isolated from various regions of Brazil and has been characterized. It is more acid tolerant and uses sucrose. Root isolates of the two other species have also been shown to be more tolerant to acid conditions than soil isolates such as strain Sp 7. In comparison with Azotobacter, Azospirillum spp occurred in higher numbers in the rhizosphere and in higher numbers on the root surface and within the roots of various C_4 and C_3 cereals and grasses, even in alkaline soils where Azotobacter is usually found in higher numbers. Many new reports on inoculation effects in pot and field experiments give a much more solid basis for future application, especially in soils where nitrogen is limiting and where Azospirillum spp do not naturally occur in high numbers. In some specific environments, lack of establishment or survival of Azospirillum spp in the rhizosphere has been made responsible for inoculation failures.

Low or intermediate nitrogen fertilizer levels, ample water supply, and in some cases application of organic materials have been shown to enhance inoculation effects or naturally occurring N_2-fixation. The need for proper strain selection has been stressed. In general Azospirillum inoculation effects on N-incorporation by cereals were more pronounced in adult plants and on grain yields and especially nitrogen incorporation into grains as compared to vegetative parts of the plant. Inoculation effects on young plants were suggested to be mainly due to growth promoting substances rather than nitrogen fixation. In some instances inoculation effects on N-incorporation can be related to enhanced nitrogen assimilation from the soil due to a larger root system or to nitrate reduction by Azospirillum within the roots.

It is remarkable how much information has accumulated within the ten years after the rediscovery of Spirillum lipoferum (now Azospirillum) as a nitrogen fixing bacterium. Although the importance and wide distribution of this organism has been firmly established, progress towards understanding the Azospirillum plant association is slow. One of the reasons seems to be the lack of reliable methods to study this process under laboratory conditions.

An urgent need to list strains and standardize strain designations for all Azospirillum strains currently used, with their origins and characteristics, was emphasized by the participants.

LIST OF PARTICIPANTS

Balandreau, J.	Nancy	France
Bally, R.	Nancy	France
Bazzicalupo, M.	Firenze	Italy
Beck, T.	München	Germany
Beyer, D.	Bayreuth	Germany
Beyse, J.	Bayreuth	Germany
Blum, E.	Göttingen	Germany
Bothe, H.	Köln	Germany
Christiansen-Weniger, C.	Kiel	Germany
Del Gallo, M.	Roma	Italy
Döbereiner, J.	Rio de Janeiro	Brazil
El-Khawas, H.	Cairo	Egypt
Elmerich, C.	Paris	France
Fahsold, R.	Bayreuth	Germany
Großmann, K.	Limburgerhof	Germany
Handschin, G.	Basel	Switzerland
Hartmann, A.	Bayreuth	Germany
Hegazi, N.A.	Cairo	Egypt
Heinrich, D.	Stuttgart	Germany
Heulin, T.	Nancy	France
Hoffmann, R.	Bayreuth	Germany
Horn, D.	Erlangen	Germany
Jagnow, G.	Braunschweig	Germany
Kapulnik, Y.	Rehovot	Israel
Kleiner, D.	Bayreuth	Germany
Klingmüller, W.	Bayreuth	Germany
Marocco, A.	Lodi	Italy
Martin, P.	Hohenheim	Germany
Maretzki, A.	Honolulu, Hawai	U.S.A.
Menze, H.	Göttingen	Germany
Mertens, T.	Stuttgart	Germany
Morpurgo, G.	Roma	Italy
Nair, S.K.	Paris	France

Neumayr, L.	Kulmbach	Germany
Nguyen, N.D.	Bayreuth and Hanoi	Germany and Vietnam
Okon, Y.	Rehovot	Israel
Omar, N.	Nancy	France
Pedrosa, F.O.	Brighton and Curitiba	England and Brazil
Renwick, A.	Braunschweig and Durham	Germany and England
Rösch, A.	Erlangen	Germany
Schwabe, G.	Bayreuth	Germany
Singh, M.	Bayreuth	Germany
Smith, R.L.	Gainesville, Florida	U.S.A.
Stolp, H.	Bayreuth	Germany
Sundman, V.	Helsinki	Finland
Vincenzini, M.	Firenze	Italy
Weber, U.	Bayreuth	Germany
Wenzel, W.	Bayreuth	Germany
Wolfenden, I.	Lancaster	England